Results of the II National Research Project of AIAr: Archaeometric Study of the Frescoes by Saturnino Gatti and Workshop at the Church of San Panfilo in Tornimparte (AQ, Italy)

Results of the II National Research Project of AIAr: Archaeometric Study of the Frescoes by Saturnino Gatti and Workshop at the Church of San Panfilo in Tornimparte (AQ, Italy)

Editors

Anna Galli
Mauro Francesco La Russa
Maria Francesca Alberghina
Alessandro Re
Donata Magrini
Celestino Grifa
Rosina Celeste Ponterio

Basel • Beijing • Wuhan • Barcelona • Belgrade • Novi Sad • Cluj • Manchester

Editors

Anna Galli
Department of
Materials Science
University of Milano-Bicocca
Milan, Italy

Mauro Francesco La Russa
Department of Biology,
Ecology and Earth Science
University of Calabria
Arcavacata di Rende, Italy

Maria Francesca Alberghina
Department of Biology,
Ecology and Earth
Sciences (DiBEST)
University of Calabria
Arcavacata di Rende, Italy

Alessandro Re
Department of Physics
University of Turin
Turin, Italy

Donata Magrini
ISPC-CNR
Istituto di Scienze del
Patrimonio Culturale
Sesto Fiorentino, Italy

Celestino Grifa
Department of Science
and Technology
University of Sannio
di Benevento
Benevento, Italy

Rosina Celeste Ponterio
Institute for Chemical-Physical Processes
National Research Council of Italy (IPCF-CNR)
Messina, Italy

Editorial Office
MDPI
St. Alban-Anlage 66
4052 Basel, Switzerland

This is a reprint of articles from the Special Issue published online in the open access journal *Applied Sciences* (ISSN 2076-3417) (available at: https://www.mdpi.com/journal/applsci/special_issues/72762643J2).

For citation purposes, cite each article independently as indicated on the article page online and as indicated below:

Lastname, A.A.; Lastname, B.B. Article Title. *Journal Name* **Year**, *Volume Number*, Page Range.

ISBN 978-3-0365-8838-4 (Hbk)
ISBN 978-3-0365-8839-1 (PDF)
doi.org/10.3390/books978-3-0365-8839-1

Cover image courtesy of Simone Caglio.

© 2023 by the authors. Articles in this book are Open Access and distributed under the Creative Commons Attribution (CC BY) license. The book as a whole is distributed by MDPI under the terms and conditions of the Creative Commons Attribution-NonCommercial-NoDerivs (CC BY-NC-ND) license.

Contents

Anna Galli, Maria Francesca Alberghina, Alessandro Re, Donata Magrini, Celestino Grifa, Rosina Celeste Ponterio and Mauro Francesco La Russa
Special Issue: Results of the II National Research Project of AIAr: Archaeometric Study of the Frescoes by Saturnino Gatti and Workshop at the Church of San Panfilo in Tornimparte (AQ, Italy)
Reprinted from: *Appl. Sci.* 2023, *13*, 8924, doi:10.3390/app13158924 1

Letizia Bonizzoni, Simone Caglio, Anna Galli, Luca Lanteri and Claudia Pelosi
Materials and Technique: The First Look at Saturnino Gatti
Reprinted from: *Appl. Sci.* 2023, *13*, 6842, doi:10.3390/app13116842 15

Luca Lanteri, Sara Calandra, Francesca Briani, Chiara Germinario, Francesco Izzo, Sabrina Pagano, et al.
3D Photogrammetric Survey, Raking Light Photography and Mapping of Degradation Phenomena of the Early Renaissance Wall Paintings by Saturnino Gatti—Case Study of the St. Panfilo Church in Tornimparte (L'Aquila, Italy)
Reprinted from: *Appl. Sci.* 2023, *13*, 5689, doi:10.3390/app13095689 31

Letizia Bonizzoni, Simone Caglio, Anna Galli, Chiara Germinario, Francesco Izzo and Donata Magrini
Identifying Original and Restoration Materials through Spectroscopic Analyses on Saturnino Gatti Mural Paintings: How Far a Noninvasive Approach Can Go
Reprinted from: *Appl. Sci.* 2023, *13*, 6638, doi:10.3390/app13116638 47

Francesco Armetta, Dario Giuffrida, Rosina C. Ponterio, Maria Fernanda Falcon Martinez, Francesca Briani, Elena Pecchioni, et al.
Looking for the Original Materials and Evidence of Restoration at the Vault of the San Panfilo Church in Tornimparte (AQ)
Reprinted from: *Appl. Sci.* 2023, *13*, 7088, doi:10.3390/app13127088 63

Francesca Briani, Francesco Caridi, Francesco Ferella, Anna Maria Gueli, Francesca Marchegiani, Stefano Nisi, et al.
Multi-Technique Characterization of Painting Drawings of the Pictorial Cycle at the San Panfilo Church in Tornimparte (AQ)
Reprinted from: *Appl. Sci.* 2023, *13*, 6492, doi:10.3390/app13116492 79

Luigi Germinario, Lorena C. Giannossa, Marco Lezzerini, Annarosa Mangone, Claudio Mazzoli, Stefano Pagnotta, et al.
Petrographic and Chemical Characterization of the Frescoes by Saturnino Gatti (Central Italy, 15th Century)
Reprinted from: *Appl. Sci.* 2023, *13*, 7223, doi:10.3390/app13127223 97

Valeria Comite, Andrea Bergomi, Chiara Andrea Lombardi, Mattia Borelli and Paola Fermo
Characterization of Soluble Salts on the Frescoes by Saturnino Gatti in the Church of San Panfilo in Villagrande di Tornimparte (L'Aquila)
Reprinted from: *Appl. Sci.* 2023, *13*, 6623, doi:10.3390/app13116623 117

Alessia Andreotti, Francesca Caterina Izzo and Ilaria Bonaduce
Archaeometric Study of the Mural Paintings by Saturnino Gatti and Workshop in the Church of San Panfilo, Tornimparte (AQ): The Study of Organic Materials in Original and Restored Areas
Reprinted from: *Appl. Sci.* 2023, *13*, 7153, doi:10.3390/app13127153 133

Silvia Ferrarese, Davide Bertoni, Alessio Golzio, Luca Lanteri, Claudia Pelosi and Alessandro Re
Indoor Microclimate Analysis of the San Panfilo Church in Tornimparte, Italy
Reprinted from: *Appl. Sci.* **2023**, *13*, 6770, doi:10.3390/app13116770 . . . . **147**

Sara Calandra, Irene Centauro, Stefano Laureti, Marco Ricci, Teresa Salvatici and Stefano Sfarra
Application of Sonic, Hygrometric Tests and Infrared Thermography for Diagnostic Investigations of Wall Paintings in St. Panfilo's Church
Reprinted from: *Appl. Sci.* **2023**, *13*, 7026, doi:10.3390/app13127026 . . . . **165**

Editorial

Special Issue: Results of the II National Research Project of AIAr: Archaeometric Study of the Frescoes by Saturnino Gatti and Workshop at the Church of San Panfilo in Tornimparte (AQ, Italy)

Anna Galli [1], Maria Francesca Alberghina [2,*], Alessandro Re [3], Donata Magrini [4], Celestino Grifa [5], Rosina Celeste Ponterio [6] and Mauro Francesco La Russa [2]

1. Department of Materials Science, University of Milano-Bicocca, Via R. Cozzi 55, 20125 Milan, Italy; anna.galli@unimib.it
2. Department of Biology, Ecology and Earth Sciences (DiBEST), University of Calabria, 87036 Arcavacata di Rende, Italy; mauro.larussa@unical.it
3. Department of Physics, University of Torino, Via Pietro Giuria 1, 10125 Turin, Italy; alessandro.re@unito.it
4. Institute of Heritage Science, National Research Council (ISPC-CNR), Area della Ricerca Firenze, Via Madonna del Piano 10, 50019 Sesto Fiorentino, Italy; donata.magrini@cnr.it
5. Department of Science and Technology, University of Sannio in Benevento, Via De Sanctis snc, 82100 Benevento, Italy; celgrifa@unisannio.it
6. Institute for Chemical-Physical Processes, National Research Council (IPCF-CNR), V. le F. Stagno d'Alcontres 37, 98158 Messina, Italy; ponterio@ipcf.cnr.it
* Correspondence: francesca.alberghina@gmail.com

1. The AIAr Project

The archaeometric study of the frescoes by the painter Saturnino Gatti (1463–1518) in the apse of the Church of San Panfilo in Villagrande di Tornimparte (L'Aquila) was the subject of the II National Research Project [1] conducted by members of the Italian Association of Archaeometry (AIAr). The research activities were carried out as part of a scientific agreement of the AIAr, signed in 2020 by the Abruzzo Regional Secretariat of the Ministry for Culture and Superintendence of Archaeology, Fine Arts and Landscape for the provinces of L'Aquila and Teramo. Several non-destructive in situ investigations and laboratory analyses on micro-fragments sampled from the different levels of the pictorial cycle were carried out thanks to the co-working of 21 Research Groups, with more than 60 AIAr researchers involved in the different stages of scientific studies of pictorial materials and of the environmental conditions in which the frescoes are now preserved.

The present Special Issue is an important opportunity to illustrate for the first time the results of the pre-restoration diagnostic study. The research project for the archaeometric study of the cycle of frescoes by the painter Saturnino Gatti (Figure 1) was carried out through the application of analytical methodologies made available by the Research Groups as members of the AIAr. The technical–scientific study of the pictorial cycle at the church of San Panfilo was aimed at providing useful indications both from a purely cognitive point of view, as a deepening of the artistic technique of the painter and his collaborators, and from a conservative point of view. In fact, the knowledge of the materials and the evaluation of the state of conservation are essential to supporting and guiding the methodological approach for the future planning of restoration works of the frescoes, helping to guarantee the principles of sustainability, durability and compatibility of the materials on which a correct conservative approach is based.

(a)   (b)

Figure 1. (**a**) Church of San Panfilo in Villagrande di Tornimparte in the province of L'Aquila. (**b**) Cycle of frescoes in the apse by Saturnino Gatti (1463–1518) object of study and research activities.

The scientific investigation campaign, whose methods and results are described in detail in the articles of this Special Issue, was identified following the technical inspection carried out by the restorers and art historians. This preliminary phase was aimed at understanding the needs in terms of knowledge and conservation features, the extension of the areas to be investigated for the documentation of the degradation and the management methods of the in situ and laboratory investigations.

The macro-objectives followed by the working groups coordinated by the Scientific Board of AIAr were:

- to document the state of conservation of the architectural building and the painted surfaces;
- to analyse the microclimate and the level of inertia of the church in relation to external conditions;
- to understand the degradation phenomena of the pictorial surfaces and the masonries;
- to identify and map materials of previous restoration works superimposed onto the original pictorial surfaces;
- to characterize the original materials and pictorial techniques ("a fresco" vs. "a secco");
- to identify the artistic technique and typical features of Saturnino Gatti and the other painters who worked in San Panfilo church.

In detail, the activities and the objectives addressed after a fruitful dialogue with staff delegated to the conservation and safeguarding of the cycle of frescoes are those illustrated below:

A. Digital photogrammetric surveys aimed at the documentation of the state of conservation of the architectural artefact and the paintings. The fully non-invasive technique guarantees the robust restitution of three-dimensional geometries and a high-resolution rendering of the pictorial surface. The obtained photorealistic 3D digital model can be used not only for mere documentation purposes, but also as a measurable support for diagnostic mapping. Moreover, the 3D model, deriving from a replicable photographic process, will allow researchers to quickly and effectively document the pre- and post-intervention.

B. Microclimate monitoring and analysis of the level of inertia of the building with respect to external conditions. The acquisition, processing and analysis of thermo-hygrometric values for the assessment of environmental conditions allow researchers to: (i) detect the condition of the thermo-hygrometric parameters, temperature (T, °C) and relative humidity (RH, %); (ii) compare the measured values with the microclimatic parameters for the correct conservation of the artefacts of different nature identified by the Recommendation [2,3]; (iii) analyse the daily ranges of temperature and relative humidity and compare them with the values indicated by the Recommendation [2,3] in order to evaluate the influence of the conservation conditions;

(iv) calculate the dew point temperature values for highlighting the condensation phenomena; and (v) indicate any corrective measures (active or passive) for reaching the thermo-hygrometric parameters for correct conservation.

C. Mapping of the restored areas and identification of the restoration materials that overlapped during past documented and undocumented conservation interventions. Through the integration of diagnostic imaging techniques and spectrometric analyses, the areas affected by repainting, remakes, stuccos, and protective materials applied to the original surfaces have been localized in detail. The characterization of these overlapping materials provided a guide to assess applicable treatments and the removal of materials that compromise the durability of the original layers. Particular attention was paid to understanding the blue background layer on the vault surface, heavily darkened, to evaluate what the original appearance was and how many layers and which materials were superimposed over time (Figure 2).

Figure 2. Details of the vault affected by a dark blue background and comparison between the current conditions and the ICR pre-intervention status in a historical image (**bottom right**), taken from Archivio storico e Archivio fotografico della Soprintendenza Archeologia Belle Arti e Paesaggio di L'Aquila e Teramo.

In fact, the site has undergone various restoration interventions over time, not all of which have been documented, that have variously attempted to address the conservation issues and have had aesthetic results that were not always respectful to the formal and stylistic reading of the original figurative complex (Figure 3). At a first macroscopic evaluation, these interventions were faced with different criteria from time to time, generating a palimpsest that needs to be reconstructed for a correct diachronic and stratigraphic interpretation of the remakes in relation to the original layers still present under the repaintings. The interpretation of the diagnostic results constitutes objective support for evaluating the removal of areas of renovation which currently partially or totally hide traces of the Renaissance frescos.

Figure 3. Examples of typical alterations found in the lower and upper portions of the apsidal basin.

D. Characterization of the original materials of the "a fresco" area and any "a secco" layers, including both inorganic (pigments and plasters) and organic (binders, additives, etc.) constituents. The identification of the palette in the various scenes of the pictorial cycle could allow researchers to evaluate the presence of different "interventions" or periods, providing art historians with further data for an objective identification of the portions attributable to the Saturnino Gatti workshop, to date hypothesized only based on historical–stylistic features.

E. Study Saturnino Gatti's artistic technique and the organization rules of his workshop by obtaining information on the techniques of transporting the drawing and the construction of the figurative system, in an attempt to identify operational peculiarities that could distinguish portions painted by Saturnino Gatti from those painted by collaborators. During the preliminary inspection, it was possible to observe the complex system of engravings, particularly in some exposed areas due to the partial or total lacunae of the pictorial layer (Figure 4). Furthermore, in a close view of the surfaces, the typical dots of the "spolvero" technique are visible.

Figure 4. Examples of the several engravings visible both in the pictorial portions characterized by total collapse and in those with a good state of conservation.

This first evidence was deepened during the in situ photographic and multispectral investigation. This diagnostic approach based on multispectral imaging also allows for a better iconographic interpretation, highlighting the volumetric effects of the figures and the perspective relationships altered by degradation phenomena or repainting that has overlapped over time. Furthermore, in order to obtain a deeper identification of artist technique and understanding of the technical choices of the workshop, the study envisaged the characterization of the plasters (number of layers, type of aggregates, binders, presence of primers, etc.) through mineral–petrographic analyses.

F. Documentation, mapping and understanding of the phenomena of degradation (current or pre-existing) of the pictorial surfaces and the architectural structures on which they stand (Figure 5). This phase of the study returned a lot of fundamental information to document the different types of degradation, to guide the design choices and to guarantee the durability of the future restoration intervention. In fact, suitable analytical methodologies were used to locate the pictorial or wall areas affected by water infiltration from windows or roofing and capillary rising damp.

Figure 5. Details of the alterations observed at the different levels of the pictorial surfaces: (**a**,**b**) details of the pictorial surfaces altered along the apse; (**c**,**d**) details of the alterations affecting the pictorial surfaces of the vault.

Analyses of soluble salts which generate efflorescence (and/or sub-fluorescence) were also conducted. The identification of salts allowed for understanding the degradation phenomena in progress that involve the constituent materials or external compounds introduced by the surface migration process, in order to evaluate the appropriate removal treatments. The study of the cross-section evaluated the degradation effects generated by crystallization and solubilization on the original portions and on the integration of mortars implemented during the numerous previous interventions applied to the site during the 20th century, while also considering the external and internal climatic conditions. With regard to the plasters and the pictorial layer, compositional and stratigraphic investigations (in situ and in-lab) were carried out aimed at understanding the phenomena of chromatic alteration or selective detachments, both in relation to current or past conservation conditions and to the peculiarities of the artistic technique.

To achieve these objectives, the study was performed, in the first phase, via in situ/non-invasive techniques using portable instrumentation and, successively, by means of micro-destructive analyses on samples collected from accessible surfaces. The integrated multi-analytical approach was selected due to the complementarity of the expected results as well as minimizing the number of samples, guaranteeing as much as possible the integrity of the surfaces analysed.

The non-invasive analyses, performed on the pictorial layers of the whole apse, involved: (1) photogrammetric surveys based on computer vision algorithms integrated with SfM (Structure from Motion); (2) measurement and monitoring of thermos-hygrometric parameters; (3) multispectral investigations—infrared reflectography (IRR), false colour infrared (FC-IR), imaging of luminescence in the visible induced by ultraviolet (UVL) lighting; (4) hypercolorimetric multispectral imaging (HMI); (5) digital optical microscopy in visible and ultraviolet light; (6) traditional and/or pulse compression IR thermography (PuCT, pulse-compression thermography); (7) X-ray fluorescence spectrometry (XRF); (8) external reflectance Fourier transform infrared spectroscopy (ER-FTIR) and/or total attenuated reflectance (ATR-FTIR); and (9) Raman spectroscopy.

On the other hand, the micro-destructive techniques on fragments selected on the basis of the first results of the in-situ investigations concerned: (1) Raman spectroscopy and surface-enhanced Raman scattering (SERS); (2) polarizing optical microscopy (OM) and scanning electron microscopy with microanalysis (SEM-EDS) on thin and polished sections; (3) ion chromatography for the analysis of soluble salts; (4) X-ray diffraction analysis (XRD); and (5) mass spectrometry analyses.

Figure 6 showed the sampling areas of the 47 micro-fragments sampled from the apsidal basin and on the central part of the vault. In particular, the five pictorial panels were been identified with the letters A, B, C, D, E as follows:

- Panel A—first painting from the left of the apse basin, depicting "Christ in the Garden of Gethsemane";
- Panel B—depicting the "Coronation with Thorns";
- Panel C—depicting the "Crucifixion" (no longer visible due to deterioration and previous interventions);
- Panel D—fourth panel from the left of the apsidal basin, depicting the "Lamentation over the Dead Christ";
- Panel E—fifth panel from the left of the apse basin—Resurrection;
- Vault—central part depicting "God the Father Blessing".

Figure 6. Panels (from left of the apse) (**A–E**) and (**vault**): Location and denomination of the areas where the 47 fragments of plaster and pictorial layers were taken. Each sampling areas have been selected on the basis of the first results of the in-situ investigations.

The following section describes the history of the building and the most relevant historical aspects about the frescoes (patronage, chronology, iconographic programme, and artistic reference models), information useful for the understanding and contextualization of the results provided by the project.

2. Saturnino Gatti's Frescoes in Their Historical Context by Saverio Ricci (Ministero della Cultura, Soprintendenza Archeologia, Belle Arti e Paesaggio per le Province di L'Aquila e Teramo)

The church of San Panfilo in Villagrande (declared as a National Monument in 1902) is the only one, among the others located in the several small villages called "ville" belonging to the municipality of Tornimparte, that is still preserved in its original architectural shape. Founded around the year 1000, the church is preceded by a large, raised arcade inside of which are walled ashlars from the Roman Era and other fragments from the Middle Ages, including a stone that is engraved with the date 1471—in the opinion of some historians, this could be proof that the church was partially rebuilt after the earthquake of 1461. This belief was expressed by some authors [4,5], according to whom over the centuries the building was extensively reconstructed in those parts that had fallen because of historical earthquakes (among them, the Great Earthquake of 1703 and the Marsica Earthquake of 1915). The façade is bordered on the left by a sloping wall and on the opposite side by the bell tower; it appears to be of modest elevation due to its disproportionate width, having to cover the space of four naves, which are also flanked on the left side by another longitudinal room, practically the same size as the side nave, formerly occupied by the ossuary and used since 1832 as the Oratory of the Congregation of Our Lady of Sorrows. The part above the arcade was fully renovated in the 20th century, as can be seen by observing the rose window, no longer adorned with theories of small arches framing an orbicular light, but simply closed by a modern stained-glass window depicting the titular saint of the church, Saint Pamphilus, a long-time evangeliser, bishop, and patron of Sulmona who lived in the 7th century. The interior space is divided into four naves: two to the right, the central nave, and the left one. In the left aisle is the Chapel of the Crucifix and two altars dedicated to St Francis and St Pamphilus. In the first right aisle is the altar of the Nativity, while in the adjacent aisle are the chapel of the Visitation and altars dedicated to the Holy Rosary and St Joseph.

In any case, the building's fame is mainly due to the cycle decorating the apse, which was commissioned by the Community to the painter Saturnino di Giovanni di Gatto, born in the nearby village of San Vittorino (nowadays in the municipality of Pizzoli). It has been possible to establish the chronology of the cycle in a rather precise manner, since the archival findings delimit with a good margin of precision the interval within which Saturnino Gatti executed the decoration of the apse. The first document is dated 23 May 1489, a payment mandate with which Saturnino received forty-five florins from "Dominico Antoni Paulutii de Tornamparte" for the frescoes to be painted in an unspecified chapel in the church; on 1 May 1490, the commissioned artist signed the contract to paint the apse of the same church; on 19 April 1491, the "massari" of Tornimparte procured the money to be paid to Saturnino for the work in progress, renting the pastures of Villagrande for three seasons; the last balance for the paintings is dated 12 December 1494, but already in February, two renowned masters, Silvestro dell'Aquila and Sebastiano di Cola da Casentino, had been called "ad pretiandum picturam", i.e., summoned to the site as experts to estimate the completed works in the "cappella sive tribuna in dicta ecclesia" [6]. Some historians have argued that these were two separate commissions, but the most complete transcription and interpretation of the documents carried out by [5,7] seems to have removed any doubts in this regard, demonstrating that the documents transcribed are related to the same decorative enterprise: the payment received in 1489 should be interpreted as an advance payment, and the work was carried out from 1491 to 1494, as also stated by Gatti's most authoritative scholar, Ferdinando Bologna [8].

Turning to the topic of narrative structure and subjects illustrated in the cycle, note how the apsidal dome hosts a representation of Paradise with the imposing figure of God the Father Blessing at the centre of the vault, surrounded by the apostles and male saints (on his right), the Virgin Mary and female saints (on the left side), a whirlwind of Cherubs arranged around the perimeter of the mandorla within which God the Father stands out, angels in flight scattering flowers and a choir of fifteen musician angels, as evidenced by a scroll containing a short fragment of a musical score supported by two angels with ribbons, which reads "Gloria in excelsis Deo". The Gregorian chant evoked belongs to an ancient mass, dating back to the 11th century, widely used in the Catholic liturgy [9]. The apsidal basin originally featured the following scenes from the Passion of Christ divided into five panels, depicting in the following order, counter clockwise: (1) Capture of Christ in the Garden of Gethsemane; (2) Coronation with Thorns; (3) Crucifixion (?); (4) Lamentation over the Dead Christ; (5) Resurrection.

Of the original cycle, the first, fourth and fifth panels remain intact, while the second was completely repainted, including the faux architecture framing it, probably in the second half of the 17th century. The third panel, on the other hand, is almost entirely lost, also due to the unfortunate opening, in uncertain times, of a large window to allow light into the room. It is interesting to observe the iconographic complexity of the first panel, which hosts two other episodes secondary to the main scene, closely connected to it but chronologically to be fixed before the moment in which the Capture of Christ takes place. These are: (a) the Kiss of Judas, in the background and characterised by the presence, never noted before, of a turreted city seen from a distant point of view, identifiable as Jerusalem, concealed by a thick veil due both to the Saturnino-esque use of aerial perspective (and thus the desire to artificially recreate the effect of atmospheric haze), and to the accumulation of particles on the paint film; (b) St. Peter amputating Malchus' ear—in this case, Saturnino placed the figures in the foreground, although the episode was narrated as contemporary with the betrayal perpetrated by Judas, thus preceding the moment when Jesus, after having healed the servant of the High Priest mutilated by the apostle, hands himself over to the Sanhedrin guards, as told in the Gospel of St. John.

The Stories of the Passion are anticipated, on the left side, by a mock niche within which stands, with a harsh and angular physicality and in a sharply characterized posture, Saint Vitus the martyr accompanied by two dogs, whose presence is explained by the great popular devotion paid to the thaumaturge saint, especially in the territories with a prevalent pastoral economy (it is no coincidence that Abruzzo, Molise and Apulia, regions of transhumance, are still strongholds of his cult). Given that shepherds lived in close proximity with domestic animals, the saint was invoked against dog rabies and the bite of poisonous animals. It should also be pointed out that in Colle San Vito, a hamlet very close to Villagrande, inside the church consecrated to St Vitus is conserved a wooden statue of the saint attributed to Saturnino Gatti in recent contributions [10,11]. Two Doctors of the Church, St. Jerome and St. Ambrose, are portrayed in the window gaps behind the first panel, while above the opening, in the upper intrados, is painted the famous "IHS" Christogram inscribed in a sun (an invention of St. Bernardine of Siena), very popular in the Abruzzo Apennines due to the enormous veneration that the Franciscan preacher enjoyed at an early age in L'Aquila. Also note the detail that develops from the base of the left side niche: at this point, there was a trompe-l'oeil-painted carpet unrolled downwards, of which unfortunately only the yellow-coloured base and some faint preparatory traces of the arabesque weave design remain visible. Its presence recalls a customary tradition of Renaissance painting, especially in the Adriatic area, between Venice and the Marches (oriental carpets were imported in the country via the ports on the Adriatic Sea).

On the right-hand side, the cycle is concluded by another mock niche containing the figure of Fra Pietro dell'Aquila, a Franciscan theologian and philosopher native to Tornimparte, better known by the Latin pseudonym "Scotellus" (it. Scotello). This inclusion may have different explanations: the affection of the inhabitants towards their illustrious fellow citizen, to whom the Community—the commissioner of the works—wanted to pay visible homage. In another hypothetical scenario, Scotello's commentaries on the Holy Scriptures were, probably, the source of inspiration for the cycle, or guided the selection of the peculiar episodes of the Passion of Christ that formed the basis of the iconographic programme. A third hypothesis, finally, leads us to interpret the choice as a precise symbolic intention, the dichotomy between Faith and Reason, identifying in Saint Vitus the emblem of an irrational credence bordering superstition, and in Scotello the simulacrum of humanist thought.

The arch above the high altar is decorated with eight figures of prophets looking out from false niches that illusionistically open in the surface of the subarch, while on the extrados of the same triumphal arch we find a large representation of the Annunciation, surrounded by urban scenery that can be interpreted as a view of L'Aquila, in which the primordial Basilica of San Bernardino da Siena is clearly recognisable: the imposing dome, the highest ever built in Abruzzo at that time, was in fact completed in 1489 and therefore was emphatically depicted by Gatti just a few years later [12]. All the paintings are framed by faux perspective architecture, elements of exceptional artistic value, designed and composed according to the Renaissance tradition: some of the best-preserved details are the Corinthian capitals, particularly that corresponding to the access door of the sacristy. In addition, a plinth of mock mottled marble slabs starting from the floor, above which rise grotesque decorated pillars surmounted by a continuous entablature, creates the overall effect of a series of openings onto external landscapes that break through two-dimensional space.

To conclude with a brief stylistic overview of the cycle, it is necessary to premise that the research has not yet fully shed light on the training of the artist from San Vittorino. Modern critics believe, almost unanimously, that Gatti, after an initial frequentation of Silvestro dell'Aquila's workshop (in 1477, when he was only fourteen years old, he appears as a witness in the stipulation of a deed with the executors of the will of Cardinal Amico Agnifili, concerning the allocation of the ecclesiastic's funerary monument in the Cathedral of San Massimo), in the following decade the artist stayed in Florence. Indeed, no activity is documented in the L'Aquila archives until 1488, when he undertook to paint a chapel in the church of San Domenico. For many reasons, his style appears to be increasingly linked to the teachings of Andrea del Verrocchio, although in the past, some art historians suggested recognising the influence of other great Renaissance artists such as Melozzo da Forlì, Antoniazzo Romano and Pietro Perugino. I totally agree with the opinions of Ferdinando Bologna [13], Alessandro Angelini [14] and Michele Maccherini [7,15], who have dedicated some important studies, with very convincing insights, to that season of "verrocchismo aquilano" developed in the last thirty years of the 15th century, in which the Florentine extraction of Gatti's art has been emphasized when comparing certain details of the Tornimparte frescoes with Verrocchio's artworks. Saturnino, in the end, can rightly be defined as the most faithful interpreter of his master in Abruzzo. But this matter should be faced by scrupulously verifying derivations and affinities between Verrocchio's and Gatti's preparatory drawings, sculptures, and paintings, while also including various other Tuscan (and non-Tuscan) artists who, in the same period of Gatti's apprenticeship, received their artistic rudiments in Verrocchio's workshop, often perpetuating his expressive solutions, compositional ideas and executive techniques. Unfortunately, few and incomplete examples of Verrocchio's fresco paintings remain (essentially the detached fresco of San Domenico in Pistoia), which makes it more interesting to understand how much Saturnino's work is imbued with the technical notions he learnt alongside his master. For example, the use of calcium white spread even on partially dry plaster is very notable (Figure 7), and it is remarkable that it was also found in Luca Signorelli's frescoes in the Abbey of Monte Oliveto Maggiore (1497–1498) and at the Chapel of San Brizio in Orvieto's Cathedral (1499–1502), practically contemporary with those of Tornimparte.

Among the figures in the cycle more characterized for physiognomic data, postures, and expressiveness (and therefore most like Verrocchio's repertoire), angels, the risen Christ and sleeping soldiers in the Resurrection panel, the dead Christ and pious women in the Lamentation panel, Jesus and the apostles in the Capture of Christ panel can be identified. The saints in the vault and prophets in the sub-arch must also originally have shown much more refined features (see as an example Figure 8), appearing closer than today to the models of Verrocchio's school. Investigations have unequivocally shown, for example, that the extensive repainting of the sky in the Paradise background was carried out with Prussian Blue, which pertains to a modern restoration [16].

Figure 7. A comparison between Mary Magdalene in the Lamentation panel (**left**) and the detached fresco Madonna with standing Child (L'Aquila, church of Santa Margherita) (**right**) reveals identical executive technique in highlights made in "bianco di calce" (calcium white) for the finest details like the veils, and traces of gold leaf used for the gilding of halos and edges of robes.

Figure 8. Prophet Daniel appearing in the sub-arch reveals, through the comparison of pictures dating back from 1920 to 2020 (from left), the almost total loss of the "a secco" finishes located in the incarnate, eyes, hair and in the scroll held in the left hand.

Other retouches almost certainly took place at the same time, for which zinc white was used and therefore cannot be dated before the mid-19th century [17]. Another interesting aspect that emerges from the surveys is the possibility of recognising and consequently preserving the finishing touches applied "a secco": the petrographic and chemical studies have demonstrated the tendency of Saturnino to apply colour corrections onto the dry plaster [18].

During the restoration phase, the cleaning should be approached with particular care and caution to remedy the serious errors of evaluation that were repeated in the cleaning operations executed in the 1950s and 1972 [19], because of which many exquisite details were removed due to the use of excessively invasive products. All these arguments, however, deserve further investigation when the planned restoration work is actually launched.

3. Results

The monitoring of the thermo-hygrometric conditions within the site was the very first step. The activity, organized into two main phases (intensive and continuous measurements), started in February 2021 and ended in April 2022 [20]. The microclimate characterisation inside the church, the comparison between average monthly indoor and outdoor temperatures and the application of the EN 15757 standard [2,3] supply the restorers with information regarding the range of historical variability of relative humidity that should be respected during the restoration project.

With a view to supporting the work of the restorers, 3D photogrammetry, raking light documentation and mapping of the degradation phenomena affecting the painted surfaces of the apsidal conch were performed [21]. This fundamental step supplied relevant information on the conservation conditions, on the execution techniques and on the materials, but also helped in the selection of the points for sampling to perform subsequent laboratory analysis.

As far as the state of conservation of the painted surfaces is concerned, the decorated apsidal conch of St. Panfilo Church shows a significant overlap of damage pathologies probably attributable to both the executive techniques and the microclimate conditions of the site for which more detailed information can be found in many of the contributions in this Special Issue. Calandra et al. [22], by applying the sonic pulse velocity test, hygrometric tests, and infrared thermography, give some clues about the state of conservation of the frescoes, i.e., the combined system of plaster and wall support. The complete analysis of the frescoes' state of conservation revealed several areas of detachment or degradation phenomena. Thanks to the use of two analytical techniques (IC and ATR-FTIR) performed on micro-samples [23], it was possible to characterize the degradation due to salt crystallization. Specifically, efflorescence was mainly caused by newly formed crystals of gypsum and calcium carbonate, mainly due to the capillary rise from the ground, with the exception of the samples taken close to the window splays and from the vault, in which infiltration could be responsible. The bad state of conservation of these paintings could also be due to previous invasive restorations that affected the surfaces with heavy retouching and repainting. This is especially observed close to the openings of panel A and panel C, where modern mortar-based materials have been used, probably leading to the occurrence of the observed degradation phenomena [18]. After the investigations centred in the characterisation of the degradation phenomena, a diagnostic study was conducted aimed at the characterization of the executive techniques over the time, with particular attention being paid to the preparatory drawings, the original materials of the draft fresco and the occurrence of "a secco" finishes still present, both inorganic (pigments and plasters) and organic (binders and additives in the plasters or in the dry application to define the details of the representations).

The RAK photos [21] as well as the images acquired with imaging techniques [17] show the presence of incision marks (with minor pouncing) and the boundaries of "giornate" and "pontate", which clearly suggest the use of the fresco painting technique, although in several areas the colour spread was "a secco" [19,24,25], especially where colour loss is observed. The presence of a "morellone" layer under the blue-sky background for the scene depicted on panel E and under the vault confirms the use of the "a secco" painting technique [16].

The strong synergy between the in situ/non-invasive analyses [16] and the imaging techniques survey [17,21] performed on the same pictorial areas allowed an analysis of the quality of the punctual analyses results, demonstrating how a well-designed non-invasive campaign can drastically reduce the amount of sampling [18,19,24,25] necessary to obtain complete information on the materials and the artist's technique.

The palette of Saturnino Gatti comprises pigments, either used in their pure form or in a mixture, to create different hues that are all compatible with the coeval pictorial technique. Moreover, the non-invasive approach suggests the use of a gilding technique with a golden leaf adhered to a red bolus preparation in areas where it is no longer visible to the naked eye. Moreover, in addition to the original palette, the analytical protocol identified modern paint materials, such as Prussian blue, zinc white, copper, and chromium-based greens, used for numerous retouching. These results have been confirmed by petrographic and chemical analyses performed on microsamples.

The analyses of binders and restoration products revealed a variety of organic materials on the mural paintings, most of which are from past restoration interventions and have a synthetic origin. The overspread presence of paraffin is likely due to the application of a mineral wax-based coating/consolidant. In particular, the execution technique encompassed the use of tempera-based paints, while retouched areas are characterised by the presence of oil-based resins [25].

The multi analytical approach presented in this Special Issue provided useful indications about both the materials and the executive techniques of the painter Saturnino Gatti and his collaborators, both to guide the methodological choices during the imminent planning of the restoration intervention.

Author Contributions: Conceptualization, M.F.L.R., M.F.A. and A.G.; writing—original draft preparation, M.F.A. and A.G.; writing—review and editing, M.F.L.R., M.F.A., A.G., C.G., A.R., R.C.P. and D.M.; supervision M.F.L., M.F.A. and A.G.; project administration, C.G.; funding acquisition, M.F. and L.R. All authors have read and agreed to the published version of the manuscript.

Funding: The project activities were financed thanks to the funds allocated within the scientific agreement stipulated by A.I.Ar and the MIBACT Regional Secretariat for Abruzzo (ref. protocol MIBACT_SR-ABR | 10/11/2020 | 0003539-P).

Acknowledgments: We would like to thank the art historian Saverio Ricci (saverio.ricci@cultura.gov.it) for the research conducted in: Archivio storico e Archivio fotografico della Soprintendenza Archeologia Belle Arti e Paesaggio di L'Aquila e Teramo; Archivio Centrale dello Stato di Roma. The use of the archive documents and pictures in this Special Issue was made possible by the kind permission from the Ministero della Cultura. Last but not least we want to thank the Proloco Tornimparte association and the mayor of the municipality of Tornimparte for their availability and support shown during the diagnostic campaign.

Conflicts of Interest: The authors declare no conflict of interest.

References

1. Available online: https://www.associazioneaiar.com/wp/blog/risultati-progetto-saturnino-gatti/ (accessed on 4 September 2022).
2. UNI 10829; Works of Art of Historical Importance—Ambient Conditions for the Conservation—Measurements and Analysis. UNI Standard Ente Nazionale Italiano di Unificazione: Milano, Italy, 1999.
3. EN 15757; Conservation of Cultural Property-Specifications for Temperature and Relative Humidity to Limit Climate-Induced Mechanical Damage in Organic Hygroscopic Materials. UNE-EN, AENOR: Madrid, Spain, 2010.
4. Moretti, M. *Architettura Medioevale in Abruzzo (dal VI al XVI Secolo)*; De Luca Ed.: Roma, Italy, 1971; pp. 808–809.
5. Mannetti, T.R.; Chelli, N.; Vecchioli, G. *Saturnino Gatti nella chiesa di San Panfilo a Tornimparte*; Edizioni del Gallo Cedrone: L'Aquila, Italy, 1992; pp. 11–12 and pp. 21–36.
6. Chini, M. Documenti relativi ai pittori che operarono in Aquila fra il 1450 e il 1550 circa. *Bull. Della Regia Deput. Abruzz. Stor. Patria S. 3* **1927**, *XVIII*, 13–138, 63.
7. Maccherini, M.; Pezzuto, L. *Saturnino Gatti e la sua Bottega in L'arte Aquilana del Rinascimento, a cura di M*; Maccherini, L'Una Ed: L'Aquila, Italy, 2010; pp. 121–154.
8. Bologna, F. *Saturnino Gatti: Pittore e Scultore nel Rinascimento Aquilano*; Textus: L'Aquila, Italy, 2014; p. 132.
9. Arbace, L. I volti dell'anima. In *Saturnino Gatti: Vita e opere di un artista del Rinascimento*; Paolo De Siena Editore: Pescara, Italy, 2015; p. 71.
10. Monopoli, R. Una Proposta per Saturnino Gatti: Il San Vito Ligneo di Tornimparte. Master's Thesis, Università degli Studi dell'Aquila, L'Aquila, Italy, 2010/2011.
11. Principi, L. Un San Sebastiano di Silvestro dell'Aquila e un San Vito di Saturnino Gatti. In *Il Capitale Culturale*; Department of Education, Cultural Heritage and Tourism, University of Macerata: Macerata, Italy, 2015; Volume 11, pp. 11–39.
12. Ciranna, S. La costruzione della cupola di San Bernardino all'Aquila tra XV e XVIII secolo. In *Lo Specchio del Cielo: Forme, Significati, Tecniche e Funzioni della Cupola dal Pantheon al Novecento*; Conforti, C., Ed.; Electa: Milano, Italy, 1997; pp. 151–165. (pp. 157, 164).
13. Bologna, F. Saturnino Gatti: Un'opera. *Paragone* **1950**, *5*, 60–63.
14. Angelini, A. Saturnino Gatti e la congiuntura verrocchiesca a L'Aquila. In *I da Varano e le Arti*; De Marchi, A., Falaschi, P.L., Eds.; RI OPAC: Ripatransone, Italy, 2003; Volume 2, pp. 839–854.
15. Maccherini, M. Artisti e suggestioni toscane in Abruzzo. In *Condivisione di Affetti*; Firenze e Santo Stefano di Sessanio; Opere d'arte dalla Galleria degli Uffizi; Natali, A., Ed.; Maschietto: Firenze, Italy, 2011; pp. 27–54.
16. Bonizzoni, L.; Caglio, S.; Galli, A.; Germinario, C.; Izzo, F.; Magrini, D. Identifying Original and Restoration Materials through Spectroscopic Analyses on Saturnino Gatti Mural Paintings: How Far a Noninvasive Approach Can Go. *Appl. Sci.* **2023**, *13*, 6638. [CrossRef]
17. Bonizzoni, L.; Caglio, S.; Galli, A.; Lanteri, L.; Pelosi, C. Materials and Technique: The First Look at Saturnino Gatti. *Appl. Sci.* **2023**, *13*, 6842. [CrossRef]

18. Germinario, L.; Giannossa, L.C.; Lezzerini, M.; Mangone, A.; Mazzoli, C.; Pagnotta, S.; Spampinato, M.; Zoleo, A.; Eramo, G. Petrographic and Chemical Characterization of the Frescoes by Saturnino Gatti (Central Italy, 15th Century). *Appl. Sci.* **2023**, *13*, 7223. [CrossRef]
19. Armetta, F.; Giuffrida, D.; Ponterio, R.C.; Martinez, M.F.F.; Briani, F.; Pecchioni, E.; Santo, A.P.; Ciaramitaro, V.C.; Saladino, M.L. Looking for the Original Materials and Evidence of Restoration at the Vault of the San Panfilo Church in Tornimparte (AQ). *Appl. Sci.* **2023**, *13*, 7088. [CrossRef]
20. Ferrarese, S.; Bertoni, D.; Golzio, A.; Lanteri, L.; Pelosi, C.; Re, A. Indoor Microclimate Analysis of the San Panfilo Church in Tornimparte, Italy. *Appl. Sci.* **2023**, *13*, 6770. [CrossRef]
21. Lanteri, L.; Calandra, S.; Briani, F.; Germinario, C.; Izzo, F.; Pagano, S.; Pelosi, C.; Santo, A.P. 3D Photogrammetric Survey, Raking Light Photography and Mapping of Degradation Phenomena of the Early Renaissance Wall Paintings by Saturnino Gatti—Case Study of the St. Panfilo Church in Tornimparte (L'Aquila, Italy). *Appl. Sci.* **2023**, *13*, 5689. [CrossRef]
22. Calandra, S.; Centauro, I.; Laureti, S.; Ricci, M.; Salvatici, T.; Sfarra, S. Application of Sonic, Hygrometric Tests and Infrared Thermography for Diagnostic Investigations of Wall Paintings in St. Panfilo's Church. *Appl. Sci.* **2023**, *13*, 7026. [CrossRef]
23. Comite, V.; Bergomi, A.; Lombardi, C.A.; Borelli, M.; Fermo, P. Characterization of Soluble Salts on the Frescoes by Saturnino Gatti in the Church of San Panfilo in Villagrande di Tornimparte (L'Aquila). *Appl. Sci.* **2023**, *13*, 6623. [CrossRef]
24. Briani, F.; Caridi, F.; Ferella, F.; Gueli, A.M.; Marchegiani, F.; Nisi, S.; Paladini, G.; Pecchioni, E.; Politi, G.; Santo, A.P.; et al. Multi-Technique Characterization of Painting Drawings of the Pictorial Cycle at the San Panfilo Church in Tornimparte (AQ). *Appl. Sci.* **2023**, *13*, 6492. [CrossRef]
25. Andreotti, A.; Izzo, F.C.; Bonaduce, I. Archaeometric Study of the Mural Paintings by Saturnino Gatti and Workshop in the Church of San Panfilo, Tornimparte (AQ): The Study of Organic Materials in Original and Restored Areas. *Appl. Sci.* **2023**, *13*, 7153. [CrossRef]

Disclaimer/Publisher's Note: The statements, opinions and data contained in all publications are solely those of the individual author(s) and contributor(s) and not of MDPI and/or the editor(s). MDPI and/or the editor(s) disclaim responsibility for any injury to people or property resulting from any ideas, methods, instructions or products referred to in the content.

Article

Materials and Technique: The First Look at Saturnino Gatti

Letizia Bonizzoni [1], Simone Caglio [2,*], Anna Galli [2], Luca Lanteri [3] and Claudia Pelosi [3]

[1] Department of Physics Aldo Pontremoli, University of Milano, Via Giovanni Celoria, 16, 20133 Milano, Italy; letizia.bonizzoni@unimi.it
[2] Department of Material Science, University of Milano Bicocca, Via Cozzi 55, 20125 Milano, Italy; anna.galli@unimib.it
[3] Department DEIM, University of Tuscia, Largo dell'Università, 01100 Viterbo, Italy; llanteri@unitus.it (L.L.); pelosi@unitus.it (C.P.)
* Correspondence: simone.caglio@unimib.it

Abstract: As part of the study project of the pictorial cycle, attributed to Saturnino Gatti, in the church of San Panfilo at Villagrande di Tornimparte (AQ), image analyses were performed in order to document the general conservation conditions of the surfaces, and to map the different painting materials to be subsequently examined using spectroscopic techniques. To acquire the images, radiation sources, ranging from ultraviolet to near infrared, were used; analyses of ultraviolet fluorescence (UVF), infrared reflectography (IRR), infrared false colors (IRFC), and optical microscopy in visible light (OM) were carried out on all the panels of the mural painting of the apsidal conch. The Hypercolorimetric Multispectral Imaging (HMI) technique was also applied in selected areas of two panels. Due to the accurate calibration system, this technique is able to obtain high-precision colorimetric and reflectance measurements, which can be repeated for proper surface monitoring. The integrated analysis of the different wavelengths' images—in particular, the ones processed in false colors—made it possible to distinguish the portions affected by retouching or repainting and to recover the legibility of some figures that showed chromatic alterations of the original pictorial layers. The IR reflectography, in addition to highlighting the portions that lost materials and were subject to non-original interventions, emphasized the presence of the underdrawing, which was detected using the *spolvero* technique. UVF photography led to a preliminary mapping of the organic and inorganic materials that exhibited characteristic induced fluorescence, such as a binder in correspondence with the original azurite painting or the wide use of white zinc in the retouched areas. The collected data made it possible to form a better iconographic interpretation. Moreover, it also enabled us to accurately select the areas to be investigated using spectroscopic analyses, both *in situ* and on micro-samples, in order to deepen our knowledge of the techniques used by the artist to create the original painting, and to detect subsequent interventions.

Keywords: UV fluorescence (UVF); IR reflectography (IRR); IR false colors (IRFC); hypercolorimetric multispectral imaging (HMI)

1. Introduction

This paper contributes to the special issue, "Results of the II National Research project of AIAr: archaeometric study of the frescoes by Saturnino Gatti and workshop at the church of San Panfilo in Tornimparte (AQ, Italy)". This special issue collates the scientific results of the II National Research Project, which was conducted by members of the Italian Association of Archaeometry (AIAr). For in-depth details on the aims of the project, see the introduction of the special issue [1].

The imaging analyses conducted on the wall paintings attributed to the Italian master Saturnino Gatti in the apse of the Church of San Panfilo (Tornimparte—AQ, Italy) were executed with two main purposes. First, after visualizing and identifying the areas affected by surface degradation, we aimed to map the different colored/pigmented areas in order to expand upon the amount of information available with regard to the materials used; this information was obtained using spectroscopic techniques [2–4]. The imaging techniques also allow us to obtain information concerning the primary technique used on the wall paintings, which provides art historians and conservators with useful information for historical reconstructions of the different phases of the artwork [5–8]. Imaging analysis has the advantage of making information accessible in a simple way, via the rendering of images and maps, which, through the use of colors, chromatic gradients, and the spatial distribution of values, allows for an immediate understanding of the data. For this reason, techniques that use exciting radiation, from near ultraviolet (365 nm) to near infrared (1000 nm), have been employed, thus enabling images to be acquired in different bands of the visible and near infrared range (400–1100 nm) [9–11].

Ultraviolet radiation can excite certain pigments and materials used during artistic production, causing them to emit light in the visible range, based on their specific composition. This phenomenon of luminescence can reveal important details about the artwork that may be barely visible, not visible at all, or visible under normal lighting conditions. Indeed, Ultraviolet Fluorescence (UVF) photography is a very useful documentation technique in the field of cultural heritage, especially when it occurs before any restoration work has taken place. This is because it allows us to obtain relevant information concerning the level of conservation that has taken place on the surfaces of the paintings, and it allows us to ascertain which materials were used; indeed, it can differentiate between original and added materials, such as those used during grouting and retouching processes. UVF photography is also useful for distinguishing between classes of materials that appear similar in visible light, but different under UV radiation. Moreover, it can also detect traces of materials that are no longer observable with the naked eye [12–15]. For these reasons, UVF photography is widely used in the field of restoration, and in general, it is used for the evaluation of the general state of conservation, with regard to artworks.

Infrared reflectography (IRR), as with UVF photography, is a non-invasive imaging analytical technique used for analysis. It is widely used in the conservation field and in the restoration of art objects [16]. This technique, based on the transparency of the different pigments when placed under infrared radiation, can reveal hidden features underneath the pictorial layer, such as underdrawings, *pentimenti*, and changes made by the artist during the painting process. The information obtained through IRR can be crucial for restorers, and it can help them make decisions concerning the conservation, restoration, or treatment of a painting; for example, infrared radiation can show possible material differences between pigments that visually appear similar, or it can highlight the stratified use of pigments with different opacities. Infrared reflectography can also provide valuable insights into the artist's working methods and creative process, reveal details of the artist's primary techniques via the underlying drawings, as well show *pentimenti* and changes in the settings of the scenes; indeed, it is possible that this technique can reveal the presence of previous versions of the painting [13,17].

By combining images acquired using infrared and visible light, it is possible to merge information deriving from different regions of the electromagnetic spectrum; thus, images are obtained in false colors which allow us to better discriminate between characteristics of the painting that would appear less accentuated compared to the response of the IRR image alone. When applied to wall paintings, infrared false color (IRFC) imaging can help with the identification of areas affected by damages such as detachments, cracking, discoloration, or delamination of the surface layers. Similarly to IRR, it can also help to identify the presence of underlying layers of paint or other materials that may have been covered by subsequent overpaintings or modifications; it has the advantage of presenting areas that appear irregular in different colors, rather than in grayscale, such is the case with reflectography [14,18].

Finally, hypercolorimetric multispectral imaging (HMI) was tested on two areas of the paintings. This multispectral imaging technique, developed and patented by the Roman society, Profilocolore, has been widely used to investigate easel paintings, but it has rarely been used on wall paintings; for this reason, taking the advantage of Tornimparte research project, HMI was applied in order to acquire two portions of the wall paintings and to post-process the obtained images. HMI is a powerful multispectral technique that, due to the image calibration procedure, allows us to obtain final images with a high level of reflectance and colorimetric precision [19]. The output after calibration consists of monochromatic images, to which several available algorithms in the processing software can be applied. These algorithms relate to the following: reflectance and chromatic comparison and mapping; normalized differences between monochromatic bands; principal component analysis; and the production of false color images (both infrared and ultraviolet) by simply combining the RGB channels with one of the IR bands or the UV bands [20].

Overall, the interpreted results concerning the different imaging techniques provided synergic information that was useful for verifying and mapping the conservative state of the painted surfaces, as well as obtaining preliminary information on the materials used. Such information will be expanded through the application of spectroscopic techniques performed directly *in situ*, in particular, X-ray fluorescence (XRF), Fourier Transform Infrared Spectroscopy (FTIR), Raman Spectroscopy, Fiber Optics Reflectance Spectroscopy in the UV–Vis–NIR Range (FORS) [21], and in laboratory analyses performed on samples [22].

2. Materials and Methods

Each imaging technique helps document and reveal specific and precious data, thus leading to an increased understanding of the painting's layout and history.

Image analyses were performed on 3 of the 5 panels of the lower part of the apse that still had painted layers (Panel A—the traitorous kiss of Judas and the capture of Christ in the Garden of Olives, Panel D—Deposition of Christ and Panel E—The Resurrection), as well as on the vault where the scene of the glory of God is represented. A detailed description of the painted scenes in the apse of St. Panfilo church is reported in the literature [23,24] and in a recently published paper [25]. The HMI analyses were instead focused on the flat areas of panels A and E. For a more extensive discussion of the site and of the project, refer to Galli et al. [1]. All the images were acquired from the ground with the aid of a tripod, exploiting different angles to obtain the details of the paintings; the light sources were positioned from time to time, as needed, supported by stands.

2.1. Diffuse Visible Light Photography (Vis)

Photography in diffuse visible light is always useful and often necessary when carrying out analyses of images; this allows both the documentation of the actual state of the surfaces at the time the analyses were executed, and a continuous comparison during the processing and post-production phases of the images that were acquired using other techniques. Diffuse visible light photography was conducted with an Olympus camera that had a 16 Mpx sensor and a 40–150 mm lens. Two halogen lamps were used as light sources.

2.2. Ultraviolet Fluorescence Photography (UVF)

For the Ultraviolet Florescence Photography, the images were obtained using a Nikon D800 digital reflex camera modified in Full Range to let the sensor grow accustomed to the electromagnetic spectrum from approximately 300 to 1000 nm. In front of the lens, two filters were applied: the filter A and the UV-IR cut filter, the spectra of which are shown in Figure 1.

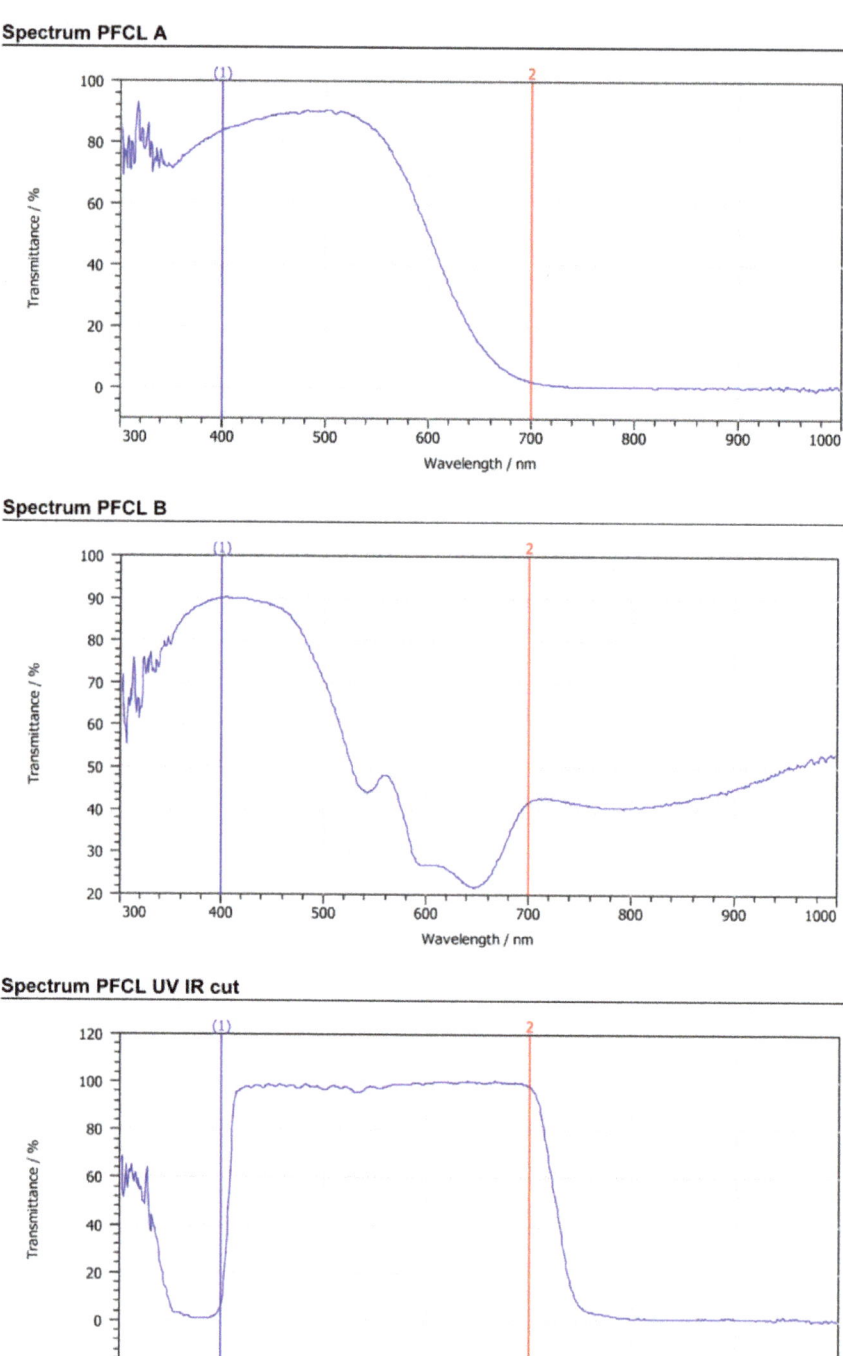

Figure 1. Spectra of cut filters, A, B, and UV-IR, used in HMI and UVF.

The UV radiation was obtained using four CR230B-HP 10W LED projectors, at peak emission at 365 nm, which were mounted at 45° in relation to the camera. LED projectors allowed us to avoid the blue-violet background that is generally obtained when using traditional UV lamps or tubes. The other advantage to using these LED projectors is that they are powered by an internal rechargeable battery; especially in the case of large surfaces, as with the case of the St. Panfilo church, they can be moved easily without the necessity of paying attention to the power supply cables or any extension cables [26].

2.3. Infrared Reflectography (IRR)

The reflectrographic infrared images were acquired using a Sony H50 digital camera with a 9 Mpx sensor and integrated optics, with an equivalent focal length of 31–465 mm. The camera was equipped with a switch that allowed us to manually remove the UV-IR cut filter placed in front of the sensor to obtain a full range in terms of sensitivity, without having to disassemble the camera. To record infrared images, an 850 nm high-pass filter was placed on the lens to limit the sensitivity range of the sensor between 850 and 1100 nm. The infrared radiation was produced by two 1000 W halogen lamps positioned at about 45° in relation to the surface in order to make the lighting homogeneous and to minimize reflections [27].

2.4. Infrared False Color (IRFC)

The Infrared False Color images were processed in two different ways: using Hypercolorimetric Multispectral Imaging (see Section 2.5) and by combining the visible and infrared images acquired with the Sony H50 camera.

After manually removing the UV-IR cut filter on the sensor of the Sony H50 camera, it was possible to acquire two (almost completely identical) pictures, with the same level of detail, in visible and infrared light. An image processing program was used to perfectly align the visible and infrared images, using some details inside the images themselves as references. Then, the channels were recombined so as to obtain the false color output image. The procedure was structured as follows: starting with the digital image, it was initially necessary to break it down into the three main channels (red, green, and blue); then, in order to replace the data contained in the channels, that which was recorded in the infrared range was inputted into the red channel, the information previously contained in the red channel was inputted into the green channel, and the information previously contained in the green channel was inputted into the blue channel. Thus, we moved from an image composed of red, green, and blue, to one composed of infrared, red, and green, by shifting the information within the three channels. [28]

2.5. Hypercolorimetric Multispectral Imaging

HMI acquisition was performed only on the flat surfaces, in accordance with a procedure described in previously published papers, and here, it was briefly summarized [29–31]. It used the same Nikon D800 digital reflex camera, modified in Full Range, during the UVF analyses.

1. The first step involves the acquisition of two images; one was acquired using filter A and the other using filter B (spectra shown in the Figure 1). The filters were screwed in front of the camera lens before each shot was taken. The lighting was set up using NEEWER (Neewer, Shenzhen, China) 750II Flashes Speedlite TTL with an LCD Display and Wireless Triggers. The flashes were modified by removing their front plastic lenses, thus allowing emissions in the 300–1000 nm region. To produce radiometrically and colorimetric calibrated images, white patches and a color-checker (36 colour samples from the Natural Color System® ©, NCS catalog) were positioned in the scene around the painting.
2. The second step concerns the calibration of the two acquired images using the proprietary software, SpectraPick® (v1.1, created by Profilocolore, Rome, Italy), that, at the end of the process, produces seven tiff files representing the multispectral

monochromatic images centered at 350 nm (UVR), 450 nm, 550 nm, 650 nm, 750 nm (IR1), 850 nm (IR2), and 960 nm (IR3), as well as the RGB 16-bit color image [32,33].

The third step of the HMI system involves the processing of the calibrated images using the software, PickViewer® (v1.0, created by Profilocolore), which provides several tools with which to obtain infrared and ultraviolet false color images, to read pixelwise colorimetry and spectral reflectance, to create similarity maps according to color or spectral data, to apply principal component analysis (PCA), and so on. The results can be saved as an image in tiff, png, or jpeg format.

2.6. Optical Microscopy in Visible Light (OM)

To document the restricted areas to be analyzed using portable spectroscopic techniques, a digital microscope with a 5 Mpx CCD sensor was also used in visible light. Using the microscope software calibration tool in conjunction with the reference target, it was possible to define a metric reference for each magnification level.

3. Results and Discussion

3.1. Material Mapping, Retouching, and Remaking

The UV fluorescence photography led to a preliminary mapping of the surfaces, which differentiated between the use of organic and inorganic materials; these presented with a characteristic fluorescence. First, it was possible to identify the presence of a significant quantity of zinc white, a pigment in use since the late eighteenth century, and therefore, associated with areas affected by pictorial retouching [33]. Indeed, the presence of zinc (Zn) was also detected by X-Ray fluorescence (XRF) in the punctual analyses of the restored areas [2]. Zinc white pigment is clearly visible due to its lemon-yellow fluorescence in the lower part of panel A, representing the betrayal of Judas. It is also visible under the window, which appears as a bright blue color in the image due to a paper which served to shield the apse from the outdoor sunlight (Figure 2) [34]. In the upper part of the same panel, especially in the background, a light yellow fluorescence is visible; this could be associated with the azurite binder, a pigment which has been recognized in the blue areas via spectroscopic analyses with XRF due to the clear presence of copper (Cu) [2]. This same fluorescence can also be observed in panel 4, the last from the left side of the apse conch, where the resurrection of Christ is represented (see Figure 3A). Areas that appear white in visible light show a faint light blue fluorescence (Figure 3B); this may be associated with calcium carbonate white, contrary to the hypothesis of the art historians that assumed that it was the use of lead white, used in the highlights of the paintings.

A detail was acquired in panel 4 wherein some fluorescent lines were observed in the general view of the wall (Figure 3A). The image in Figure 3B shows how the rays that emanate from the angels are characterized by a yellow-orange fluorescence which could be due to the presence of organic material. In accordance with art historians, the presence of possible original gilding work was hypothesized, and therefore, the organic substance whose fluorescence is observed could be referred to as the so-called *missione*, which was used to make the adhesive for the gold. Optical microscopy revealed the presence of golden traces in the rays (Figure 3C); indeed, the X Ray Fluorescence analyses, performed after the first round of image analyses, confirmed the theory that gilding had taken place. In fact, XRF spectroscopy detected the presence of gold (Au) as a chemical element that characterized those same areas [2]. The glue used for gilding was generally made of siccative oil and terpene resin; therefore, the yellow fluorescence may be associated with the lipidic component of the *missione* [35–37].

Figure 2. The UVF image of panel A, Garden of Olives.

Infrared Reflectography, together with the Infrared False Colors, allowed us to distinguish between the portions of the painting that were affected by pictorial restorations. Pigments appearing similar in visible light may exhibit a different response to infrared radiation, which is rendered at a different level of brightness in reflectographic images; this difference results in a different color response in the false color infrared images. An example is shown in panel A, in Figure 4B, in which the restoration areas in the green land are identifiable as spots with lighter levels of gray, compared with the surrounding areas. This restoration intervention is also confirmed in false color images (Figure 4C), in which the same spots exhibit a pinkish color that may be attributable to the original materials, unlike those exhibiting a green color.

Once again, via the false color images, it was possible to recognize small portions of paintings in which different materials were used. Compared with most areas, which appeared chromatically similar to the naked eye, as in the case of the blue near the Virgin's head in the scene of the deposition (Figure 5C), a single fragment shows a pink-red response in the false color image. These results are also relevant because they guided the decision concerning which points should be investigated using spectroscopic techniques [2].

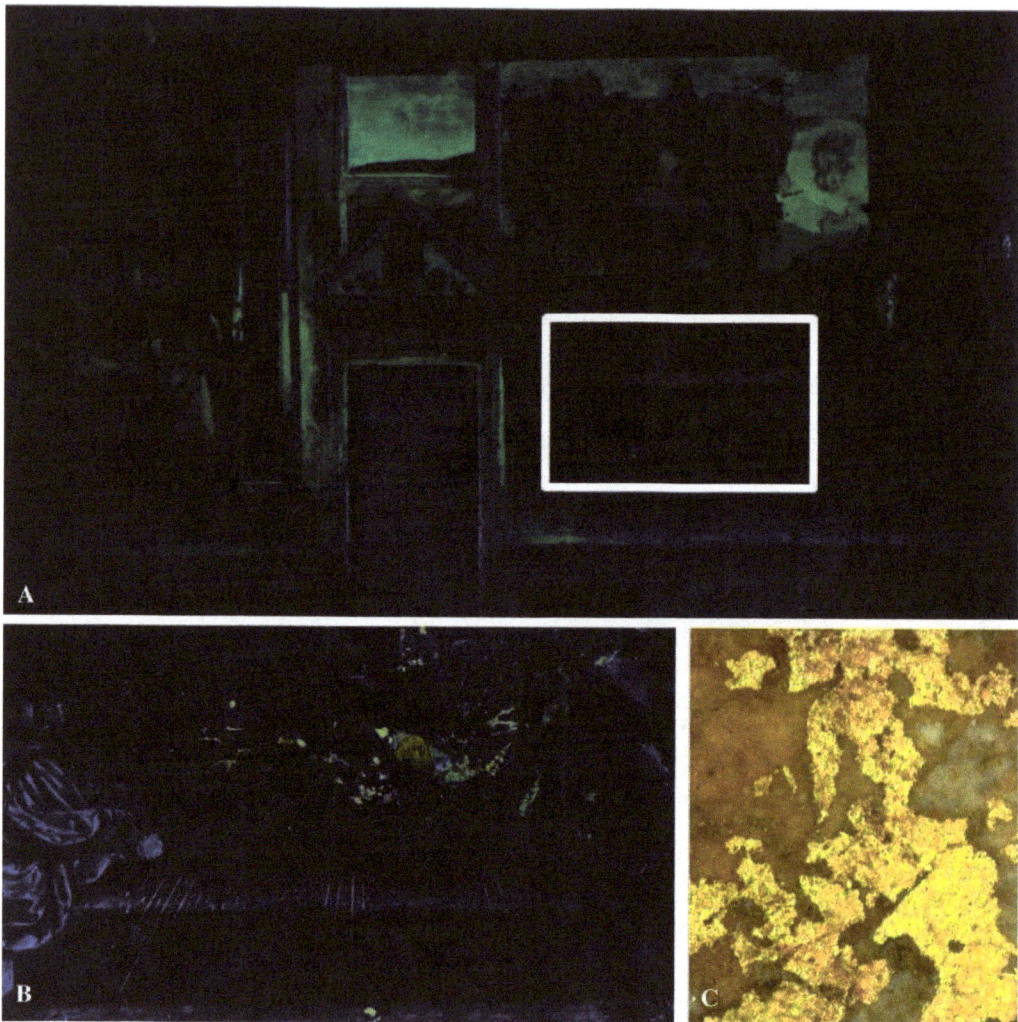

Figure 3. The UVF image of panel E, the Resurrection: (**A**) general view; the white rectangle indicates the detail shown in (**B**). (**C**) Detail of the golden traces found via OM in the rays.

Figure 4. Detail of Panel A, Garden of Olives (**A**) in visible light, (**B**) IRR, and (**C**) IRFC.

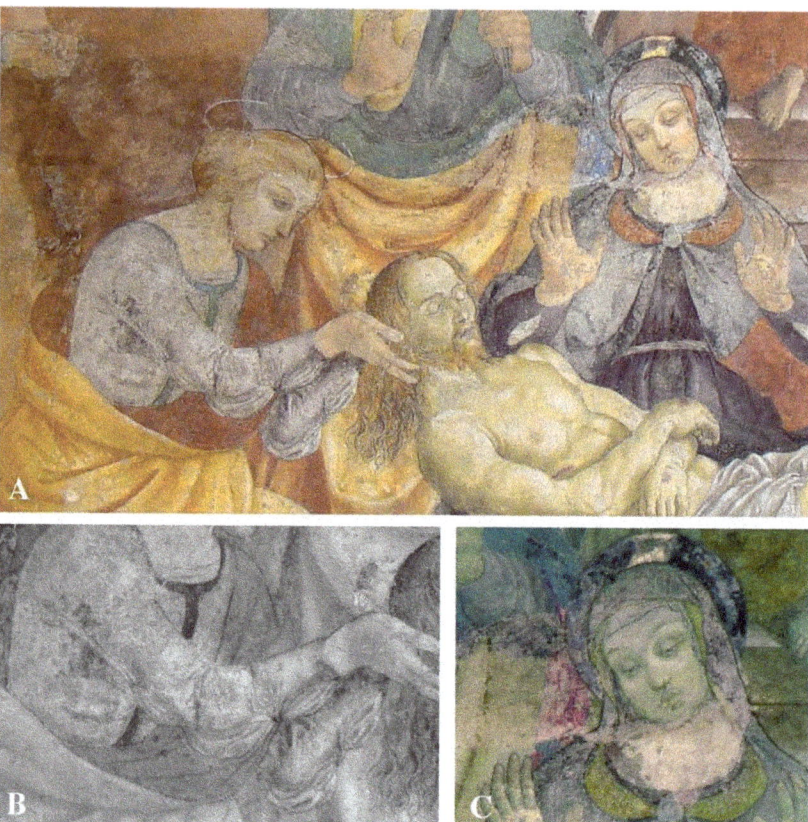

Figure 5. Detail of panel D, Deposition of Christ (**A**) in visible light, (**B**) IRR, and (**C**) IRFC.

3.2. Underdrawing

IR reflectography, in addition to allowing the mapping of the surface portions affected by material losses and highlighting the presence of restoration material, has increased the amount of information relating to artistic techniques; in particular, it proffers insight into the methods used by the artist to draw the different figures represented in the pictorial cycle [38]. Along the perimeters of the various characters, dots precisely defining the contours and the main constructive features are visible; it is known that reflectography allows us to identify underlying drawings made with materials that are opaque when examined with infrared radiation, and they are generally carbon-based [39]. The presence of the dots may therefore be due to the transfer of the drawing from cartoons or sheets via the "*spolvero*" (punching) technique. This last technique involves the creation of a full-sized drawing on a preparatory cartoon; then, the contours of the different figures are subsequently perforated with a needle or another small tip. Next, the perforated cardboard is placed on the wall's surface, and the perforated parts are dabbed with a canvas bag filled with ash, charcoal, or other fine dust that may leave a trace on the wall [40]. A clear example of such dusting is visible in image 5B, in which the series of dark dots are clearly defined near the forearm and the folds of the dress of the woman identified as Mary Magdalene [24], who supports the head of the dead Christ. Similarly, the signs of the transfer are visible on her cloak. Evidence of the same type of drawing method has been found in the other panels of the apse; this acts as evidence of a methodological unity in the realization of the entire pictorial cycle, and in support of the attribution of the artwork itself. Once again, when carefully comparing Figure 5A (visible image) with Figure 5B (reflectography), it can be seen that in some parts, such as along the shoulders, a brown pigment was used to enhance the detail of the contours, thus preventing the possibility of detecting the underlying drawing via the IRR technique.

3.3. HMI Results

To test the potential of HMI techniques, the acquired images were focused on two detailed areas of the panels A and E, as they were the only panels with flat surfaces.

The software PickViewer®, included in the HMI system, allowed us to obtain the infrared false color (IRFC) output by simply combining the RGB channels with one of the three IR channels. An example is shown in Figure 6. The retouched lacunae exhibit a pink fluorescence that better distinguishes them from the original parts. The blue dress has a dark blue response in IRFC, thus suggesting the possible presence of azurite (hypothesis confirmed via punctual analyses) [2]. In this same area, other processes were performed using the PickViewer® tools. The chromatic similarity algorithm was applied on the blue dress, as shown in Figure 7). This tool compares the values of the chromatic coordinates of the selected point (on the RGB image) with all pixels in the image, and it produces a B/W image where the white pixels have the same chromatic values as the chosen point, and the black pixels are completely different in terms of color data.

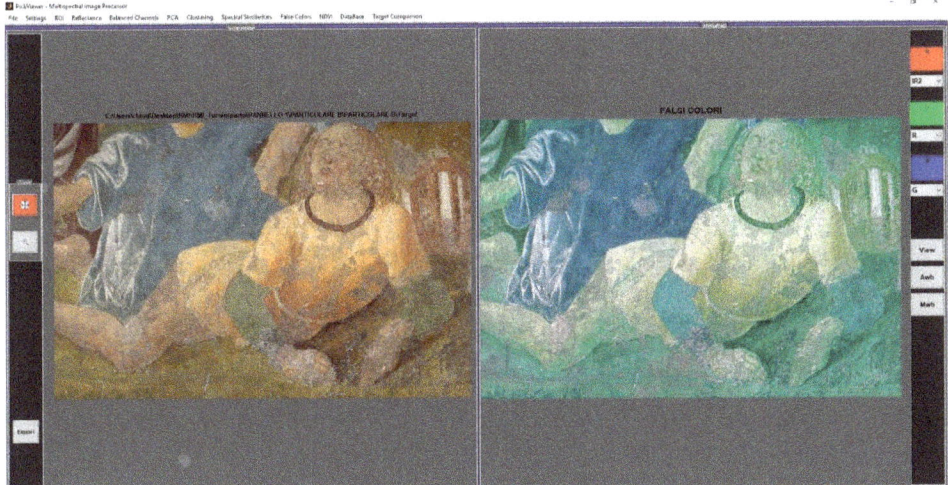

Figure 6. The graphic user interface (GUI) of the PickViewer®, shown on the **left** of the RGB image, and on the **right**, the IRFC result is shown. Detail of panel A.

In Figure 7, the black areas of the dress identify the lacunae of the painting layer or the highlights used to outline the drapery of the dress.

In the chosen area of panel A, further processes were performed, in particular, principal component analysis (PCA) was applied using the three infrared bands (IR1, IR2, and IR3). The result is shown in Figure 8, wherein the first PC highlights the drawing in the feet, in the hand, and above the sword hilt and blade (Figure 8).

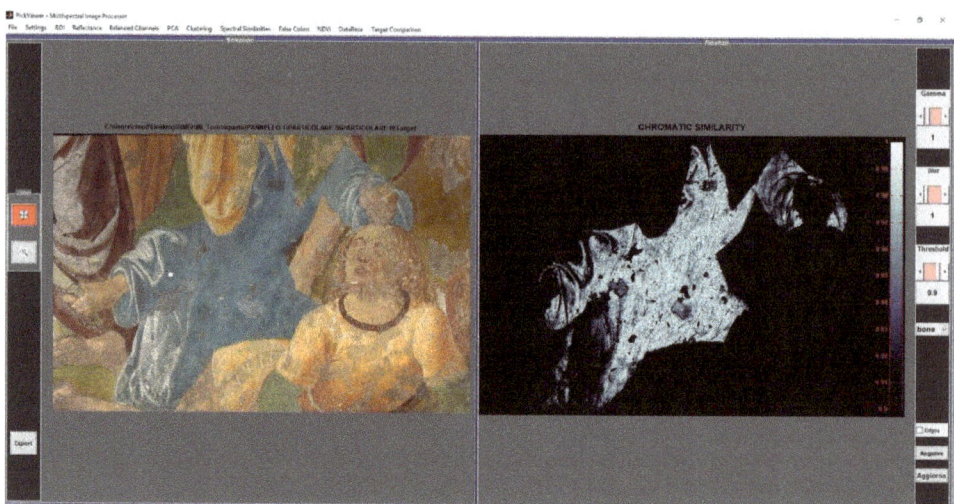

Figure 7. The GUI of the PickViewer® shows the RGB image on the **left**, with the selected point (white dot) for the application of the chromatic similarity tool, and on the **right**, the result is shown. Detail of panel A.

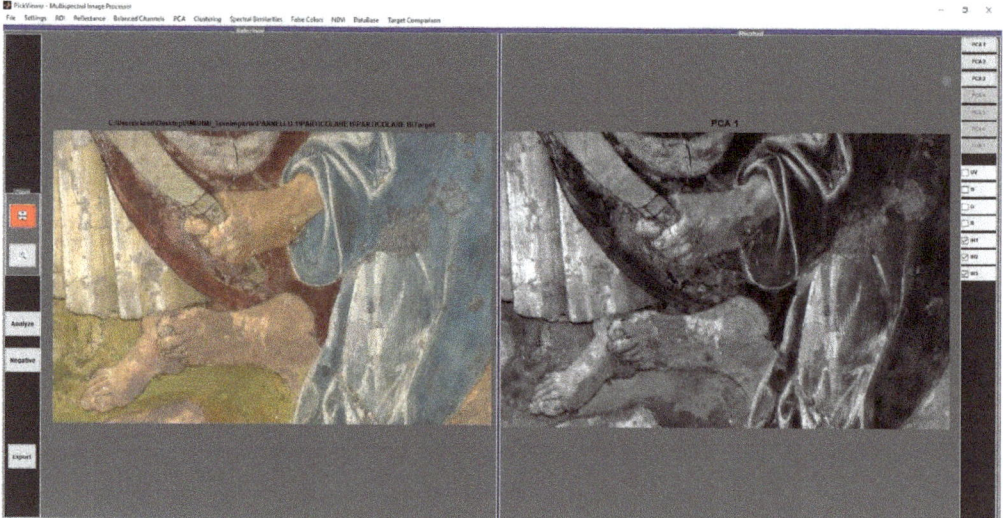

Figure 8. The GUI of the PickViewer® shows the RGB image on the **left**, and the PC1 obtained by applying the PCA to the three IR bands is shown on the **right**. Detail of panel A.

The chromatic similarity tool was also applied to a green point in panel E (Figure 9). This green, in the leg of the character that is lying down, seems to be the result of a retouch. The result of the application of the tool shows that the green retouches are located in the legs, but also in the upper part of the selected area, thus corresponding with the background. The chromatic similarity tool was also applied to a green area that seems to be the original (Figure 10).

Figure 9. The GUI of the PickViewer® shows the RGB image with the selected point (white dot on the leg of the character that is lying down, probably non-original panting) for the application of the chromatic similarity tool on the **left**, and the result is shown on the **right**. Detail of panel E.

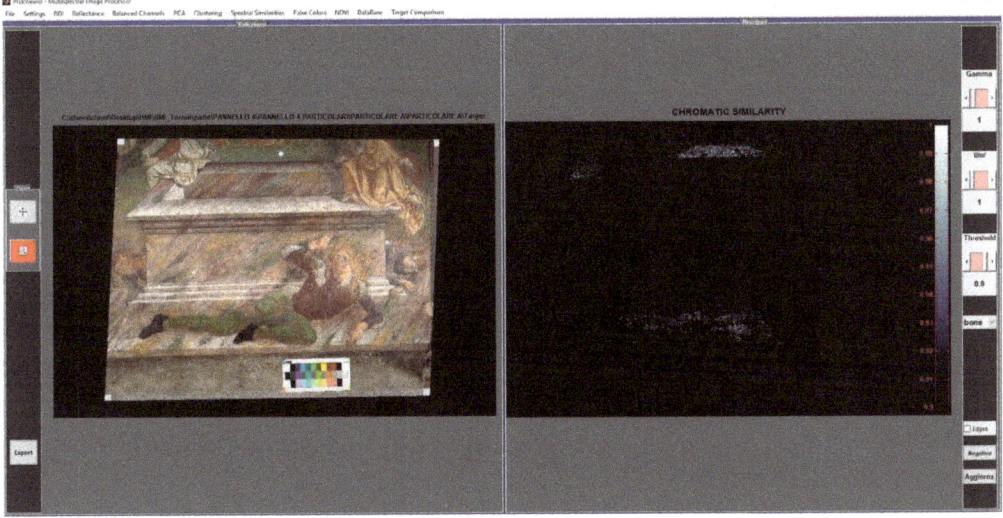

Figure 10. The GUI of the PickViewer® shows the RGB image with the selected point (white dot in the upper part of the green area, probably original) for the application of the chromatic similarity tool on the **left**, and the result is shown on the **right**. Detail of panel E.

The result of the tool's application highlights that the green in the upper part of the selected area, which corresponds with the top of the sepulcher, has a chromatic similarity with that of the legs of the lying gendarme and a small area of land on the left side, but not with the green areas that are presumed to be non-original (mapped in the Figure 9).

4. Conclusions

The data acquired in this phase of the Tornimparte project led to various results in terms of awareness, regarding the data available to progress the activities, the transition to the next steps of the investigation, and the materials and techniques used to create the wall paintings by Saturnino Gatti. First, the mapping of different pictorial materials made it possible to proceed, with greater awareness, with the decision concerning which points should be used for non-invasive spectroscopic analyses that can characterize the materials of the wall painting at the atomic and molecular level. Even the choice of micro-sampling the painting layers, which were used in the laboratory analysis, was guided by the acquired multispectral images. This approach made it possible to distinguish between the areas wherein the materials could be attributable to the original choices of the artist, and those that were subject to overpainting or restoration interventions. The use of a multispectral approach, with radiation ranging from ultraviolet to near infrared, highlighted the presence of specific materials, such as zinc white, and the recognition of several different materials in areas that appeared chromatically homogeneous.

The imaging techniques also allowed us to confirm some hypotheses concerning the executive techniques of Saturnino Gatti; the gilding, for example, was confirmed in several areas wherein the presence of visible fluorescence induced by UV radiation can be associated with the organic material used to glue the gold leaf. Small traces of the same gold leaf have been detected via digital optical microscopy at different points, such as in the rays in the risen Christ or in the halos of the saints.

In addition to the information regarding the materials of the artwork, it was also possible to provide a precise mapping of the level of conservation that had been conducted on the surfaces; it was possible to identify the presence of restoration materials, as well as the various degradation phenomena that were taking place.

Finally, as well as answering the initial research questions, it was also possible to provide information on the primary technique used to create the wall paintings; in particular, the method to transpose the drawings of the various figures to the wall. Indeed, the presence of the "spolvero" (punching) technique was unknown at the beginning of the investigation, and it provided important information on how the iconographic and scenic system was executed; in this case, a transfer technique based on the use of paper and/or cardboard was used.

Author Contributions: Investigation, L.B., S.C., A.G., C.P. and L.L.; Data Curation, L.B., S.C., A.G., C.P. and L.L.; Writing—Original Draft Preparation S.C. and C.P.; Writing—Review and Editing, L.B., S.C., A.G., C.P. and L.L.; Supervision, A.G. All authors have read and agreed to the published version of the manuscript.

Funding: This work was performed as part of Project Tornimparte—"Archeometric investigation of the pictorial cycle of Saturnino Gatti in Tornimparte (AQ, Italy)" sponsored in 2021 by the Italian Association of Archeometry AIAR (www.associazioneaiar.com).

Data Availability Statement: The data presented in this study are available on request from the Italian Association of Archeometry AIAR (www.associazioneaiar.com).

Acknowledgments: The authors want to thank the Proloco Tornimparte association and the mayor of the municipality of Tornimparte for their availability and support shown during the diagnostic campaign.

Conflicts of Interest: The authors declare no conflict of interest.

References

1. Galli, A.; Alberghina, M.F.; Re, A.; Magrini, D.; Grifa, C.; Ponterio, R.C.; La Russa, M.F. Special Issue: Results of the II National Research project of AIAr: Archaeometric study of the frescoes by Saturnino Gatti and workshop at the church of San Panfilo in Tornimparte (AQ, Italy). *Appl. Sci.* 2023; *to be submitted.*
2. Bonizzoni, L.; Caglio, S.; Galli, A.; Germinario, C.; Izzo, F.; Magrini, D. Identifying original and restoration materials through spectroscopic analyses on Saturnino Gatti mural paintings: How far a non-invasive approach can go. *Appl. Sci.* 2023, *13*, 6638. [CrossRef]

3. Bersani, D.; Berzioli, M.; Caglio, S.; Casoli, A.; Lottici, P.P.; Medeghini, L.; Poldi, G.; Zannini, P. An integrated multi-analytical approach to the study of the dome wall paintings by Correggio in Parma cathedral. *Microchem. J.* **2014**, *114*, 80–88. [CrossRef]
4. Ludwig, N.; Orsilli, J.; Bonizzoni, L.; Gargano, M. UV-IR Image Enhancement for Mapping Restorations Applied on an Egyptian Coffin of the XXI Dynasty. *Archaeol. Anthropol. Sci.* **2019**, *11*, 6841–6850. [CrossRef]
5. Gavrilov, D.; Maev, R.G.; Almond, D.P. A review of imaging methods in analysis of works of art: Thermographic imaging method in art analysis. *Can. J. Phys.* **2014**, *92*, 341–364. [CrossRef]
6. Eom, T.H.; Lee, H.S. A Study on the Diagnosis Technology for Conservation Status of Painting Cultural Heritage Using Digital Image Analysis Program. *Heritage* **2023**, *6*, 1839–1855. [CrossRef]
7. Piroddi, L.; Abu Zeid, N.; Calcina, S.V.; Capizzi, P.; Capozzoli, L.; Catapano, I.; Cozzolino, M.; D'Amico, S.; Lasaponara, R.; Tapete, D. Imaging Cultural Heritage at Different Scales: Part I, the Micro-Scale (Manufacts). *Remote Sens.* **2023**, *15*, 2586. [CrossRef]
8. Jones, C.; Duffy, C.; Gibson, A.; Terras, M. Understanding multispectral imaging of cultural heritage: Determining best practice in MSI analysis of historical artefacts. *J. Cult. Herit.* **2020**, *45*, 339–350. [CrossRef]
9. Aldovrandi, D.; Cetica, M.; Matteini, M. Multispectral image processing of paintings. *Stud. Conserv.* **1988**, *33*, 154–159.
10. Casini, A.; Lotti, F.; Picollo, M. Imaging Spectroscopy for the Non-invasive Investigation of Paintings. In *International Trends in Optics and Photonics*; Asakura, T., Ed.; Springer Series in Optical Sciences; Springer: Berlin/Heidelberg, Germany, 1999; Volume 74. [CrossRef]
11. Picollo, M.; Cucci, C.; Casini, A.; Stefani, L. Hyper-Spectral Imaging Technique in the Cultural Heritage Field: New Possible Scenarios. *Sensors* **2020**, *20*, 2843. [CrossRef]
12. Lanteri, L.; Agresti, G.; Pelosi, C. A new practical approach for 3D documentation in ultraviolet fluorescence and infrared reflectography of polychromatic sculptures as fundamental step in restoration. *Heritage* **2019**, *2*, 207–215. [CrossRef]
13. Colantonio, C.; Lanteri, L.; Ciccola, A.; Serafini, I.; Postorino, P.; Censorii, E.; Rotari, D.; Pelosi, C. Imaging Diagnostics Coupled with Non-Invasive and Micro-Invasive Analyses for the Restoration of Ethnographic Artifacts from French Polynesia. *Heritage* **2022**, *5*, 215–232. [CrossRef]
14. Cosentino, A. Effects of Different Binders on Technical Photography and Infrared Reflectography of 54 Historical Pigments. *Int. J. Conserv. Sci.* **2015**, *6*, 287–298.
15. Lanteri, L.; Pelosi, C. 2D and 3D ultraviolet fluorescence applications on cultural heritage paintings and objects through a low-cost approach for diagnostics and documentation. In Proceedings of the Optics for Arts, Architecture, and Archaeology VIII, Munich, Germany, 8 July 2021; SPIE: Bellingham, WA, USA, 2021; p. 1178417. [CrossRef]
16. Peeters, J.; Steenackers, G.; Sfarra, S.; Legrand, S.; Ibarra-Castanedo, C.; Janssens, K.; Van der Snickt, G. IR Reflectography and Active Thermography on Artworks: The Added Value of the 1.5–3 μm Band. *Appl. Sci.* **2018**, *8*, 50. [CrossRef]
17. Van Asperen de Boer, J.R.J. Infrared reflectography: A method for the examination of paintings. *Appl. Opt.* **1968**, *7*, 1711–1714. [CrossRef] [PubMed]
18. Moon, T.; Schilling, M.R.; Thirkettle, S. A Note on the Use of False-Color Infrared Photography in Conservation. *Stud. Conserv.* **1992**, *37*, 42–52. [CrossRef]
19. Melis, M.; Babbi, A.; Miccoli, M. Development of a UV to IR extension to the standard colorimetry, based on a seven band modified DSLR camera to better characterize surfaces, tissues and fabrics. In Proceedings of the SPIE 8084, Optics for Arts, Architecture, and Archaeology III, Munich, Germany, 6 June 2011; pp. 1–11. [CrossRef]
20. Vettraino, R.; Valentini, V.; Pogliani, P.; Ricci, M.; Laureti, S.; Calvelli, S.; Zito, R.; Lanteri, L.; Pelosi, C. Multi-Technique Approach by Traditional and Innovative Methodologies to Support the Restoration of a Wall Painting from the 16th Century at Palazzo Gallo in Bagnaia, Viterbo, Central Italy. *Buildings* **2023**, *13*, 783. [CrossRef]
21. Pelagotti, A.; Mastio, A.D.; Rosa, A.D.; Piva, A. Multispectral imaging of paintings. *IEEE Signal Process. Mag.* **2008**, *25*, 27–36. [CrossRef]
22. Briani, F.; Caridi, F.; Ferella, F.; Gueli, A.M.; Marchegiani, F.; Nisi, S.; Paladini, G.; Pecchioni, E.; Politi, G.; Santo, A.P.; et al. Multi-technique characterization of painting drawings of the pictorial cycle at the San Panfilo Church in Tornimparte (AQ). *Appl. Sci.* **2023**, *13*, 6492. [CrossRef]
23. Arbace, L.; Di Paolo, G. *I Volti Dell'anima, Saturnino Gatti: Vita e Opere Di Un Artista Del Rinascimento*; De Siena Editore, Ed.; De Siena: Pescara, Italy, 2012; ISBN 8896341116.
24. Mannetti, T.R.; Chelli, N.; Vecchioli, G. *Saturnino Gatti Nella Chiesa di San Panfilo a Tornimparte*; Cedrone, E.d.G., Ed.; L'Aquila Publishing: L'Aquila, Italy, 1992.
25. Lanteri, L.; Calandra, S.; Briani, F.; Germinario, C.; Izzo, F.; Pagano, S.; Pelosi, C.; Santo, A.P. 3D Photogrammetric Survey, Raking Light Photography and Mapping of Degradation Phenomena of the Early Renaissance Wall Paintings by Saturnino Gatti—Case Study of the St. Panfilo Church in Tornimparte (L'Aquila, Italy). *Appl. Sci.* **2023**, *13*, 5689. [CrossRef]
26. Groppi, F.; Vigliotti, D.; Lanteri, L.; Agresti, G.; Casoli, A.; Laureti, S.; Ricci, M.; Pelosi, C. Advanced documentation methodologies combined with multi-analytical approach for the preservation and restoration of 18th century architectural decorative elements at Palazzo Nuzzi in Orte (Central Italy). *Int. J. Conserv. Sci.* **2021**, *12*, 921–934.
27. Poldi, G.; Villa, G.C.F. *Riflettografia: Esempi Applicativi, in Dalla Conservazione Alla Storia Dell'arte*; Edizioni della Normale: Pisa, Italy, 2006; pp. 69–126.
28. Poldi, G.; Villa, G.C.F. *Infrarosso in Falso Colore, in Dalla Conservazione Alla Storia Dell'arte*; Edizioni della Normale: Pisa, Italy, 2006; pp. 127–138.

29. Laureti, S.; Colantonio, C.; Burrascano, P.; Melis, M.; Calabrò, G.; Malekmohammadi, H.; Sfarra, S.; Ricci, M.; Pelosi, C. Development of integrated innovative techniques for the examination of paintings: The case studies of The Resurrection of Christ attributed to Andrea Mantegna and the Crucifixion of Viterbo attributed to Michelangelo's workshop. *J. Cult. Herit.* **2019**, *40*, 1–16. [CrossRef]
30. Ricci, M.; Laureti, S.; Malekmohammadi, H.; Sfarra, S.; Lanteri, L.; Colantonio, C.; Calabrò, G.; Pelosi, C. Surface and Interface Investigation of a 15th Century Wall Painting Using Multispectral Imaging and Pulse-Compression Infrared Thermography. *Coatings* **2021**, *11*, 546. [CrossRef]
31. Annarilli, S.; Casoli, A.; Colantonio, C.; Lanteri, L.; Marseglia, A.; Pelosi, C.; Sottile, S. A Multi-Instrument Analysis of the Late 16th Canvas Painting, "Coronation of the Virgin with the Saints Ambrose and Jerome", Attributed to the Tuscany-Umbria Area to Support the Possibility of Bio-Cleaning Using a Bacteria-Based System. *Heritage* **2022**, *5*, 2904–2921. [CrossRef]
32. Melis, M.; Miccoli, M.; Quarta, D. Multispectral hypercolorimetry and automatic guided pigment identification: Some masterpieces case studies. In Proceedings of the SPIE 8790, Optics for Arts, Architecture, and Archaeology IV, Munich, Germany, 30 May 2013; Pezzati, L., Targowski, P., Eds.; SPIE: Bellingham, WA, USA, 2013; Volume 33, pp. 1–14.
33. Kühn, H. "Zinc White". In *Artists' Pigments: A Handbook of Their History and Characteristics*; Feller, R.L., Ed.; National Gallery of Art in association with Archetype Publications Ltd.: London, UK, 2012; Volume 1, pp. 169–186.
34. Melis, M.; Miccoli, M. Trasformazione evoluzionistica di una fotocamera reflex digitale in un sofisticato strumento per misure fotometriche e colorimetriche. In *Proc. Colore e Colorimetria Contributi Multidisciplinari*; Rossi, M., Siniscalco, A., Eds.; Maggioli Editore: Santarcangelo di Romagna, Italy, 2013; Volume IXA, pp. 28–38.
35. Sandu, I.C.A.; de Sá, M.H.; Costa Pereira, M. Ancient 'gilded' art objects from European cultural heritage: A review on different scales of characterization. *Surf. Interface Anal.* **2011**, *43*, 1134–1151. [CrossRef]
36. Darque-Ceretti, E.; Felder, E.; Aucouturier, M. Gilding of cultural heritage artefacts: An elaborated technology. *Surf. Eng.* **2013**, *29*, 146–152. [CrossRef]
37. Brocchieri, J.; Scialla, E.; Manzone, A.; Graziano, G.O.; D'Onofrio, A.; Sabbarese, C. An analytical characterization of different gilding techniques on artworks from the Royal Palace (Caserta, Italy). *J. Cult. Herit.* **2022**, *57*, 213–225. [CrossRef]
38. Daffara, C.; Fontana, R. Multispectral infrared reflectography to differentiate features in paintings. *Microsc. Microanal.* **2011**, *17*, 691–695. [CrossRef]
39. Melada, J.; Gargano, M.; Ludwig, N. Pulsed thermography and infrared reflectography: Comparative results for underdrawing visualization in paintings. *Appl. Opt.* **2022**, *61*, E33–E38. [CrossRef]
40. Chaban, A.; Tserevelakis, G.J.; Klironomou, E.; Fontana, R.; Zacharakis, G.; Striova, J. Revealing Underdrawings in Wall Paintings of Complex Stratigraphy with a Novel Reflectance Photoacoustic Imaging Prototype. *J. Imaging* **2021**, *7*, 250. [CrossRef]

Disclaimer/Publisher's Note: The statements, opinions and data contained in all publications are solely those of the individual author(s) and contributor(s) and not of MDPI and/or the editor(s). MDPI and/or the editor(s) disclaim responsibility for any injury to people or property resulting from any ideas, methods, instructions or products referred to in the content.

Article

3D Photogrammetric Survey, Raking Light Photography and Mapping of Degradation Phenomena of the Early Renaissance Wall Paintings by Saturnino Gatti—Case Study of the St. Panfilo Church in Tornimparte (L'Aquila, Italy)

Luca Lanteri [1,†], Sara Calandra [2,3,†], Francesca Briani [4], Chiara Germinario [5], Francesco Izzo [6,7,*], Sabrina Pagano [8], Claudia Pelosi [1] and Alba Patrizia Santo [2]

1 Department of Economics, Engineering, Society and Business Organization, University of Tuscia, 01100 Viterbo, Italy; llanteri@unitus.it (L.L.); pelosi@unitus.it (C.P.)
2 Department of Earth Sciences, University of Florence, 50121 Florence, Italy; sara.calandra@unifi.it (S.C.); alba.santo@unifi.it (A.P.S.)
3 Department of Chemistry, University of Florence, 50019 Florence, Italy
4 ADARTE Snc, 50141 Florence, Italy; francescabriani@adartesnc.it
5 Department of Sciences and Technology, University of Sannio, Via De Sanctis Snc, 82100 Benevento, Italy; chiara.germinario@unisannio.it
6 Department of Earth Sciences, Environment and Resources, University of Naples Federico II, Via Cinthia, 80126 Naples, Italy
7 Center of Research on Archaeometry and Conservation Science, Via Cinthia, 21, 80126 Naples, Italy
8 Department of Science of Antiquities, Sapienza Università di Roma, Piazzale Aldo Moro 5, 00185 Rome, Italy; sabrina.pagano@uniroma1.it
* Correspondence: francesco.izzo4@unina.it
† These authors contributed equally to this work.

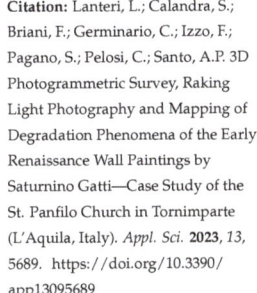

Citation: Lanteri, L.; Calandra, S.; Briani, F.; Germinario, C.; Izzo, F.; Pagano, S.; Pelosi, C.; Santo, A.P. 3D Photogrammetric Survey, Raking Light Photography and Mapping of Degradation Phenomena of the Early Renaissance Wall Paintings by Saturnino Gatti—Case Study of the St. Panfilo Church in Tornimparte (L'Aquila, Italy). *Appl. Sci.* 2023, *13*, 5689. https://doi.org/10.3390/app13095689

Academic Editor: Alexandros A. Lavdas

Received: 16 April 2023
Revised: 27 April 2023
Accepted: 3 May 2023
Published: 5 May 2023

Copyright: © 2023 by the authors. Licensee MDPI, Basel, Switzerland. This article is an open access article distributed under the terms and conditions of the Creative Commons Attribution (CC BY) license (https://creativecommons.org/licenses/by/4.0/).

Abstract: This paper provides the results of a 3D photogrammetric survey of the apsidal conch of St. Panfilo Church in Tornimparte (L'Aquila, Italy). The images were acquired and then processed in order to obtain a three-dimensional model available on Sketchfab platform. The five panels and the vault of the apsidal conch were exported from the 3D model as orthomosaics and then imported into CAD software for the mapping of the main degradation phenomena. The examined surface was almost entirely covered by mural painting and restoring mortars, the latter mainly located in the lower part of the apsidal conch. The whole surface of the apsidal conch was also examined by means of raking light that enabled highlighting of conservation problems and the presence of incision marks and *giornata/pontate* boundaries indicating the use of the *fresco* painting technique. Several degradation phenomena, attributable both to the executive technique of wall painting and the microclimate conditions, could be observed. According to the overlapping of weathering forms and the material involved, most of the examined surfaces exhibited moderate to very severe degradation.

Keywords: 3D documentation; digital photogrammetry; damage analysis and mapping; mural painting techniques; raking light; Saturnino Gatti; St. Panfilo Church; Tornimparte; decay; restoration; geomaterials

1. Introduction

The work presented in this paper may be considered one of the first steps within the archaeometric project developed by the Italian Association of Archaeometry (AIAr) in the Church of St. Panfilo in Tornimparte (L'Aquila, Italy) aimed at studying the apse wall paintings attributed to the Italian master Saturnino Gatti that were produced in his workshop in a period between 1491 and 1494 [1,2]. The idea of the project was to study the conservation status, the original and superimposed materials and the degradation phenomena affecting the wall paintings with the aim of supporting the imminent restoration intervention. The project started in 2021 and followed a suitable approach starting from in situ documentation and investigation using different techniques that supplied relevant information on the conservation conditions, on the execution techniques and on the materials, but also helped in the selection of the points for sampling to perform subsequent laboratory analysis. Documentation of this National Monument is, in fact, the first and fundamental step required to correctly perform the restoration [3–9].

With a view to supporting the work of the restorers, 3D photogrammetry, raking light documentation and mapping of the degradation phenomena affecting the painted surfaces of the apsidal conch were performed.

The standard UNI EN 16095 [10] states the importance of collecting any information concerning the state of conservation of a monument. A two-dimensional system based on the acquisition of photographs, maps and textural notes represents a common approach for the recording of sampling points, alteration phenomena and the retouching of artworks and architectural elements [11]. On the other hand, further advantages can be provided using three-dimensional models [12–14], permitting, for example, the observation of all information in a single file in which the 3D model can be easily rotated and explored. The method of 3D photogrammetry can be a valid and useful approach to obtain a 3D model that can be used as a working basis for inserting points of analysis, cleaning test areas, undertaking restoration interventions, mapping any previous interventions and, finally, viewing the paintings on platforms for sharing three-dimensional models, such as Sketchfab.

Part of the technical photography documentation can be obtained by means of raking light photography (RAK). This method allows the examination of monuments and archaeological materials, including, in particular, the observation of surface features (e.g., retouchings, lacunae, etc.), including those reflecting the painting technique, such as brush works, incision marks, as well as any other signs associated with the application of plaster and pigments [15].

The damage mapping of wall paintings is a fundamental and preliminary step in each restoration project [16]. This step, in fact, supports proper planning of correct interventions and enables evaluation of the time and cost of the entire restoration work [17]. Moreover, mapping of the damage and degradation forms can help in determining the possible causes of the damage itself [17,18]. A well-established quantitative damage diagnosis procedure, consisting of the mapping of the materials forming the monuments, as well as the weathering forms affecting their surface, is described by the "Natural stones and weathering" research group of the Aachen University of Technology (Germany), which has been used and/or adapted in several case studies [19–25].

In the present investigation, the degradation phenomena of Early Renaissance wall paintings by Saturnino Gatti were documented for the first time by means of a digital and multianalytical approach, the results of which can be easily represented or observed both by two-dimensional and three-dimensional models, as well as at different scales and levels of detail.

Since the assessments were performed at different times due to measures being taken to respond to the COVID-19 pandemic, the results of the 3D photogrammetry, RAK documentation and mapping of the degradation phenomena of the wall paintings in St. Panfilo Church are presented separately here and then discussed jointly at the end of the paper in order to support art historians and restorers to correctly plan the next intervention.

This paper contributes to the Special Issue "Results of the II National Research Project of AIAr: archaeometric study of the frescoes by Saturnino Gatti and workshop at the church of San Panfilo in Tornimparte (AQ, Italy)" in which the scientific results of the II National Research Project conducted by members of the Italian Association of Archaeometry (AIAr) are discussed and collected.

For in-depth details on the aims of the project, see the introduction to the Special Issue [26].

Brief Description of the Wall Paintings

The wall paintings, the objects of this study, were produced by Saturnino Gatti and his workshop in a period between 1491 and 1494 [1,2] in the apsidal conch of St. Panfilo Church. The wall paintings in the lower part consist of five panels (labeled from A to E) displaying some moments of the Passion of Jesus, whereas on the vault are depicted God the Father along with the Angels and Blesseds in Paradise (Figures 1 and 2).

Figure 1. Four views of the apsidal conch from the 3D rendering model.

The first scene on the left depicts the traitorous kiss of Judas and the capture of Christ (Panel A). In this scene, Judas, at the head of armed men, equipped with torches and lanterns, arrives in the Garden of Olives and greets Jesus with a traitorous kiss. As soon as Jesus is arrested, there is a reaction of dismay and at the same time defence on the part of the disciples: Peter, holding a sword and grabbing a servant of the high priest by the hair, is about to strike him and cut off his right ear; Jesus, with impressive calm, is moving towards Peter to stop him from his violent gesture but is stopped and arrested by the temple guards. Proceeding clockwise, we find the scene of the crowning with thorns where the soldiers, after having scourged Jesus, lead him to the palace of the governor Pilate, strip him of his clothes, put a red robe on him and a crown of thorny branches on his head, pressing it with sticks, outraging and mocking him as "King of the Jews" (Panel B). In the center of the apsidal circle, the fragmentary remains of the Crucifixion's scene are visible, which must have been missing for a long time, so much so that a window was opened on that wall and enlarged in 1922 (Panel C) ([2], p. 48). Continuing around the apse it is possible to observe the Deposition (Panel D): Christ, taken down from the cross, is on the knees of the Mother at whose side Mary Magdalene can be seen. The scene is observed on the right by the apostle John, and on the left by Giuseppe D'Arimatea who holds the nails of the Crucifixion in his hand. In the background, on the left, a slave of Giuseppe D'Arimatea can be seen holding a hammer and pincers used to detach Jesus from the cross; in the center three angels are depicted in front of the sepulchre ready to welcome the body of Christ.

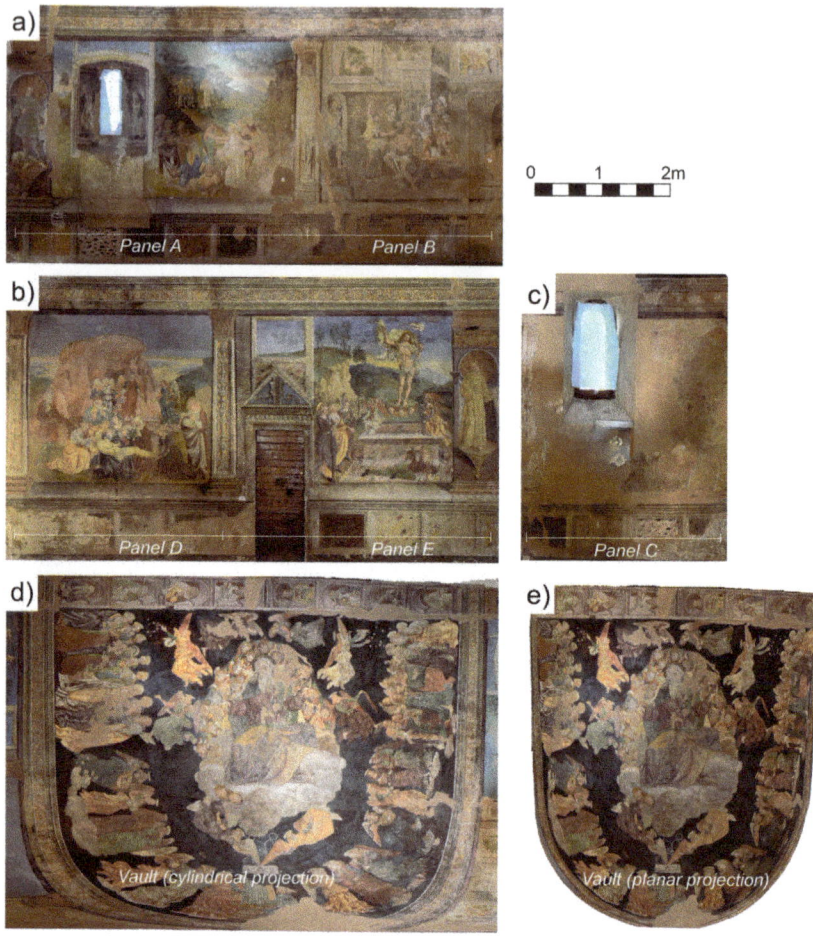

Figure 2. Orthomosaics of the apsidal conch: (**a**) panels A and B depicting the Taking and the Flagellation of Christ, respectively; (**b**) panels D and E depicting the Deposition and the Resurrection of Christ, respectively; (**c**) panel C originally depicting the Crucifixion; (**d**,**e**) cylindrical and planar projections of the vault (not to scale) depicting God the Father along with the Angels and Blessed in Paradise.

The last panel on the right side depicts the scene of the Resurrection of Christ (Panel E). In this scene, the women amazed by the empty tomb listen with amazement to the words of the angel who, sitting on the tomb with a raised forefinger, tells them that Christ is risen. One of the three guards put in charge of the sepulchre is amazed at the noise of the earthquake and at the sight of the Lord's angel, while the other two have fainted from fear. The Risen Christ, holding the banner of his triumph over death in his hand, elevated on a cloud as if supported by celestial spirits, is contemplated and adored by two angels in his resplendent glory.

The vault of the apse covers the scene of the glory of God. The Eternal Father is surrounded by angels and cherubs. On the right, the Madonna is surrounded by Holy Virgins, on the left Jesus is among the Apostles. In the background, there is Santa Cecilia among cheering angels holding musical instruments.

Just after the Resurrection, there is a niche that welcomes, under a shell, *Fra Pietro dell'Aquila*, a Franciscan friar known as *Scotellus* [1]. On the opposite side, mirroring the friar,

St. Vito is represented with two dogs. In the splay of a small window, two extraordinary half figures of St. Jerome and St. Ambrose are painted.

Finally, between the last two scenes, namely that of the deposition and that of the resurrection, the artist has created an expedient to insert a real door in an architecture of pure fiction. Above, we observe a frame which, like a window, allows the landscape to flow from one scene to another without interruption. The small door is surmounted by a triangular tympanum with the coat of arms and surrounded by frames.

2. Experimental Section

2.1. 3D Photogrammetry

The 3D model of the apsidal conch of St. Panfilo church was obtained by applying a photogrammetric approach that required the preliminary acquisition of the necessary frames, totaling 156. To do this, a Nikon D5300 digital SLR camera (Tokyo, Japan), equipped with a 17/35 mm multifocal lens, was used. To obtain the photos, the camera was mounted on a tripod and the exposure parameters were set from time to time to ensure the correct exposure with every shot. For each frame acquisition, two white LED lamps (E27 85W CFL bulb lamps, 5500 K daylight colour) were positioned on both sides of the camera, at about 45° to the subject.

After the acquisition phase, performed on-site in the church, the 156 photographs were processed by the software Agisoft Metashape® (Saint Petersburg, Russia) that can photogrammetrically process digital images, aerial and close-range photographs to obtain 3D models with the advantage of generating photorealistic surfaces.

The used software is based on the Structure from Motion (SfM) approach that, thanks to the high degree of overlapping of the acquired photos, can cover the complete geometry of the object to be rendered. This characteristic of the approach gives it the name SfM, i.e., "structure derived from a motion sensor" [27].

The working flow, widely explained elsewhere [28], enables generation of a dense points cloud that is used to generate a polygonal model (mesh), which can be texturized to return a photorealistic 3D digital model.

For the 3D model sizing of the presbytery of San Panfilo church, 10 targets were positioned on the walls, whose coordinates (x, y, z; relative to a local geographical reference system) were measured with a Total Station model Topcon GPT 7005 with angular accuracy 5″. This system guarantees a sizing of the model achieving sub-centimeter precision. The captured frames were saved in .jpg format with 24 MP resolution, 24-bit sRGB depth; each frame size was about 6 MB. The images were processed on a workstation with the following specifications: Intel(R) Core processor (TM) i7-4770 CPU @ 3.40 GHz; Ram 32 GB; 64-bit system, Windows 10®; GeForce GTX 970 4Gb video card.2.2., equipped with constructing bits RGB.

2.2. Raking Light Photography

RAK photography is a useful and simple technique for examining works of art; it allows us to detect and document surface features, especially when we are looking for details that are not visible to the naked eye and are viewed in straight light (Figure 3). For paintings, RAK photography is used to obtain information about their state of conservation and use of preparatory techniques, to document, for instance, colour retouching and loss, as well as incisions, and to study the painting technique, as this can make clear the brushwork and layering of the paints [15]. Closer examination also reveals the junctions between *giornate*, i.e., the Italian term used to describe the amount of fresco painting that has been performed in a single day, and *pontate* (*ponte*, Italian for scaffolding), in which the plaster is laid in horizontal bands.

Usually, the paintings are studied through photographs taken under diffused light conditions to avoid any reflection on the surface and to create the best lighting conditions under which the composition and colours can be better observed.

Figure 3. Detail of panel E. (**A**) image obtained in straight light; (**B**) image obtained in raking light. All the details of the picture are emphasized in (**B**) where it is possible to observe the *giornata* boundary, incisions, detachment, retouching and crack.

In the RAK technique, the paintings are enlightened in the visible band (400–780 nm) from one side only, at an oblique angle to the surface; in this way, the painted surface textures facing toward the light source are accentuated by the illumination, while those on the opposite side form shadows [15].

In this work, the raking light images were obtained through a Nikon D7000 camera, illuminating the painted surface through a LED halogen light (500W MX 500 EV) at different angles depending on the specific requirements. The entire surface of the apsidal conch was examined and photographed. The details of the painting observed through the RAK technique were described according to the literature [29].

2.3. Damage Diagnosis

The five panels (labelled from A to E) and the vault of the apsidal conch were exported as orthomosaics from the 3D model and then imported into CAD software for semi-quantitative damage diagnosis based on a monument mapping procedure of materials, weathering forms and damage categories. The weathering forms and the six damage categories were described according to the literature [19–21,29,30]. The linear damage index (DI_{lin}) and the progressive damage index (DI_{prog}) were estimated as follows:

$$DI_{lin} = [b + (c \times 2) + (d \times 3) + (e \times 4) + (f \times 5)]/100 \qquad (1)$$

$$DI_{prog} = \{[b + (c \times 4) + (d \times 9) + (e \times 16) + (f \times 25)]/100\}^{0.5} \qquad (2)$$

where b–f are the percentage areas of categories 1, 2, 3, 4 and 5, respectively. DI_{prog} and DI_{lin} range from 0 to 5 and represent, respectively, a rating of the higher damage categories and an average of them.

3. Results

3.1. 3D Photogrammetry

The very-high-resolution photorealistic rendering of the 3D spatial data generated by means of the Agisoft Metashape® software (professional education version 1.7 for universities) product directed our methodological choice, supporting a photogrammetric approach instead of use of a range-based system (e.g., TOF laser scanner). The two systems guarantee a comparable metrological precision of the measured 3D spatial data, but Metashape® has the further advantage in its workflow of creating a model rendered with a photographic resolution not achievable with the laser scanner. Furthermore, a non-secondary aspect is that the photogrammetric process requires low-cost investment when compared with range-based technologies.

This photogrammetric survey made it possible to obtain a three-dimensional, navigable and explorable model, which could also be viewed interactively in pdf format files (Figure S1).

Furthermore, the model has been uploaded to the Sketchfab platform in a non-public form, but can be made publicly visible, as carried out for other 3D models created by the University of Tuscia group [28,31]. The Sketchfab link is: https://sketchfab.com/models/88de7bbf64944408ad5fdda1d6eda141.

Moreover, Figure 1 shows additional views of the 3D model. The model can be used for mapping the state of conservation, the execution techniques, the eventual restoration areas and for georeferencing the analysis performed in the various part of the painting on the occasion of the development of the AIAr Tornimparte project.

3.2. Raking Light Photography

RAK photos of panels A, B, D, and E were taken to document the plaster work and painting technique, the preparatory drawings, and the decay phenomena of the mural painting surface. No RAK photos are available for panel C (the middle portion of the wall between panels B and D) where only a few small fragments of the mural painting have survived. With respect to this panel, the opening of a window, enlarged in 1922, has partly destroyed the mural painting with the restoration mortar covering a large part of the wall surface. Similarly, RAK photos relating to the splay window of panel A were not taken due to poor lighting and the position of the painting.

The paintings on panel A, depicting the Taking of Christ (Figure 4), display a high degree of deterioration, especially close to the window where rain infiltrated has caused localised detachments in the plaster. On the left of panel A there is a niche in which St. Vito is represented. Here, the RAK photos revealed some areas displaying colour loss and retouching (Figure 5A).

In panel A, the raking light highlighted the presence of graffiti (Figure 5B); examples of *giornate* and *pontate* boundaries are visible in Figure 5C, while colour losses and incisions can be observed in Figure 5B,D. Interestingly, in panel A, the raking light revealed the presence of a sort of landscape (Figure 6) depicting a monumental door and some religious buildings. The meaning and the origin of this detail remain unknown; we could assume a derivation from existing churches. Nevertheless, it was obviously covered by the blue sky in the background and can be classified as an example of rethinking.

Panel B (Flagellation of Christ) shows the worst state of conservation. In the original part of the painting numerous small cracks and thick brushstrokes of paint are visible as well as several incisions outlining details and figures (Figure 5E,F). The boundaries of *giornate* and the horizontal joins of *pontate* can be observed in Figure 5F.

Panels D and E (Deposition and Resurrection of Christ, respectively; Figure 4) are better preserved. The RAK photographs revealed some evident rethinking, such as the shape of the mountain in panel D (Figure 7A). In addition, *giornate* boundaries (e.g., Figure 7A,B) and several areas affected by colour loss, detachment and retouching (Figure 7C,D) were observed in panel D. The last panel on the right (panel E) showed spatula signs in the frame over the door (Figure 7D) and a layer of *morellone* below the blue sky (Figure 7E), as well as *giornate* boundaries (Figure 7E,F); retouching, colour loss and rethinking, such as the position of the tree branches (Figure 7F), were highlighted by the raking light.

Finally, to the right of panel E, another niche hosts the painting of the Franciscan friar *Fra Pietro dell'Aquila*. The painting shows numerous small cracks, loss of material and retouching.

The acquired pictures revealed that the plaster was laid inside the chapel walls by *giornate* and *pontate*, suggesting the use of *fresco* technique, with areas painted *a secco* (details about petrographic features and the microstratigraphy of the mortar-based materials are reported in [32]). Furthermore, the presence of a *morellone* layer under the blue sky background indicates the possible application of azurite with an organic binder. Finally, according to Cosentino et al. [15], the widespread presence of thick brushstokes in many scenes of the painting suggests the addition of a binder to the pigments.

Figure 4. Orthomosaics of panels A,B (**up**) and panels D,E (**down**) report the boundaries of *giornate* and *pontate*; the details of the squared areas are shown in Figures 5 and 7, respectively.

In panels A, D and E, the *giornate* boundaries generally delimit the landscape scene and the characters in the foreground; this may indicate the artist's tendency to separate the work of the landscape from the more detailed work of the figures. In contrast, in panel B the *giornate* boundaries are barely visible; this is probably due to the numerous retouchings of the painting layer.

The contour lines, often following the outline of the characters, suggest that they were made by the pressure of a pointed tool. The presence of several incisions in areas showing a total loss of colour may help to reconstruct the original drawing; these areas seem to indicate that the colour spread was *a secco*, generally considered the less durable technique over time.

Examination of the RAK photos revealed the presence of several damage effects. Significant fractures and degradation forms are evident on the entire painted surface; however, the decay forms display a more pronounced level in the different panels (panels A and B are in worse condition than the D and E panels) possibly imputable to the different environmental exposure, the position of the walls, and the presence of a window in panel A [33]. In addition, several signs of anthropic damage, especially in the form of surface scratches, are visible in the lower part of the paintings. The poor conditions of conservation necessitated restoration work and several dry retouchings over the centuries, which eventually hid some of the *giornata* boundaries, making their identification difficult in some cases.

Figure 5. Raking light photos obtained on the niche (left of panel A) and panels A and B showing retouchment (**A**), incisions (**A,D–F**), loss of colours (**A–D**), graffiti (**B**), thick brushstrokes of paint and a network of small cracks (**E**).

Figure 6. (**A**) Image of panel A; (**B**) Enlarged image of the squared area reported in A; (**C**) Raking light photo of the B area.

Figure 7. Raking light photos obtained on the panels D and E showing *giornate* and *pontate* boundaries (**A,B,E,F**), rethinking (**A,F**), network of small cracks (**C**), spatula signs (**D**), colour loss (**F**). In (**E**) is also visible a layer of *morellone* under the blue sky.

3.3. Mapping of Degradation Phenomena

The lower part of the apsidal conch consists of five panels depicting scenes from the New Testament (Figure 2). Nevertheless, only panels A, D and E show original paintings with well recognizable scenes. Conversely, the flagellation of Christ depicted on the panel C is not recognizable because of the creation of a new painting during the XVII century [1]. Lastly, the original painting of panel C, which was originally supposed to depict the Crucifixion, has completely disappeared due to extensive restoration works.

According to the geomaterial distribution map (Figure 8), most of the examined surface (ca. 87%) is covered by mural painting whereas the remaining part shows openings (2%) and restoration mortar (11%). The latter is mainly observable in the lower part of the apsidal conch, particularly in panel C.

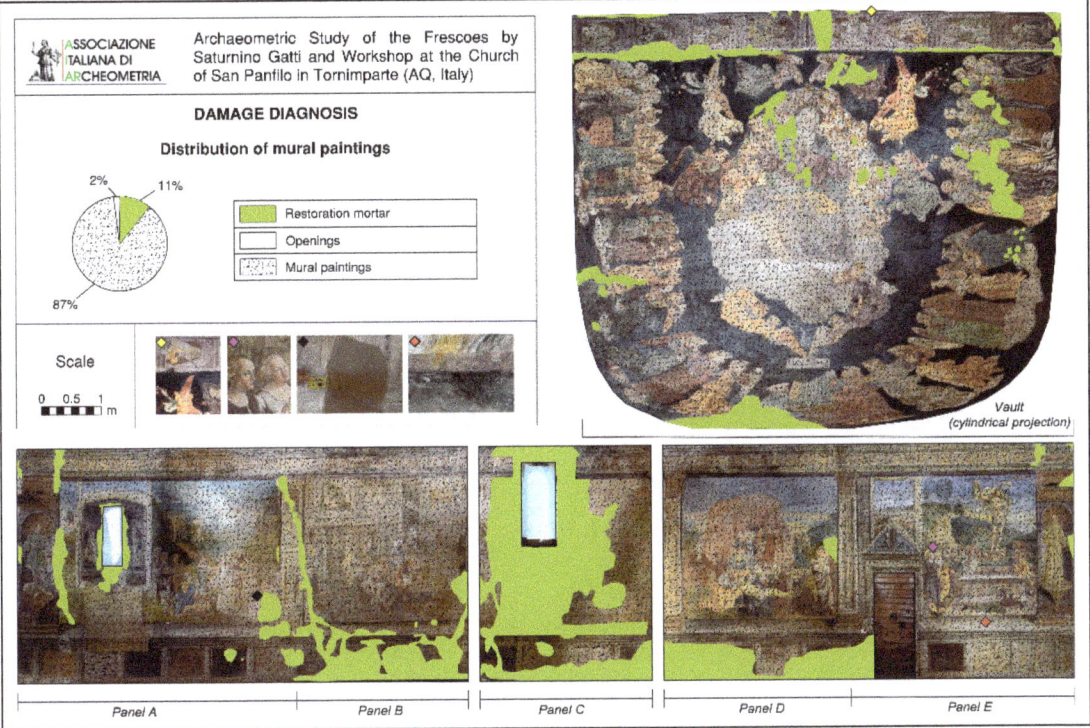

Figure 8. Distribution of geomaterials for the five panels and the vault of the apsidal conch of St. Panfilo Church in Tornimparte (AQ, Italy). Coloured dots indicate the position of the acquired pictures in the wall paintings.

Among the main weathering forms (Figures S2 and 9), individual fissures and intense networks of minor cracks (*craquele*), mainly extending on the decorated surfaces, can be observed. This kind of fissuring often evolves from minor detachments affecting the pictorial layers (Figure S2).

Frequent missing parts of variable extension, from a few millimeters (micro-*lacuna*) to several centimeters (*lacuna*), are also noticed (Figure S2). The depth of these missing parts can interrupt the pictorial layer preserving (partially) the chromatic features of the mural painting or can reach the underlying plaster exposing the pouncing and incision marks. In the lower portion of the panels, it is possible to observe frequent *graffiti* (scratching) (Figure S2).

The chromatic alteration of the wall paintings is generally due to the concomitance of discoloration phenomena (bleaching), moisture trapping and surface deposits (Figure S2), as well as colour changes that are probably due to the transformation of azurite in malachite (Figure S2) [32,34]. In the map of the weathering forms (Figure 9), the whole painted surface is considered as a precaution.

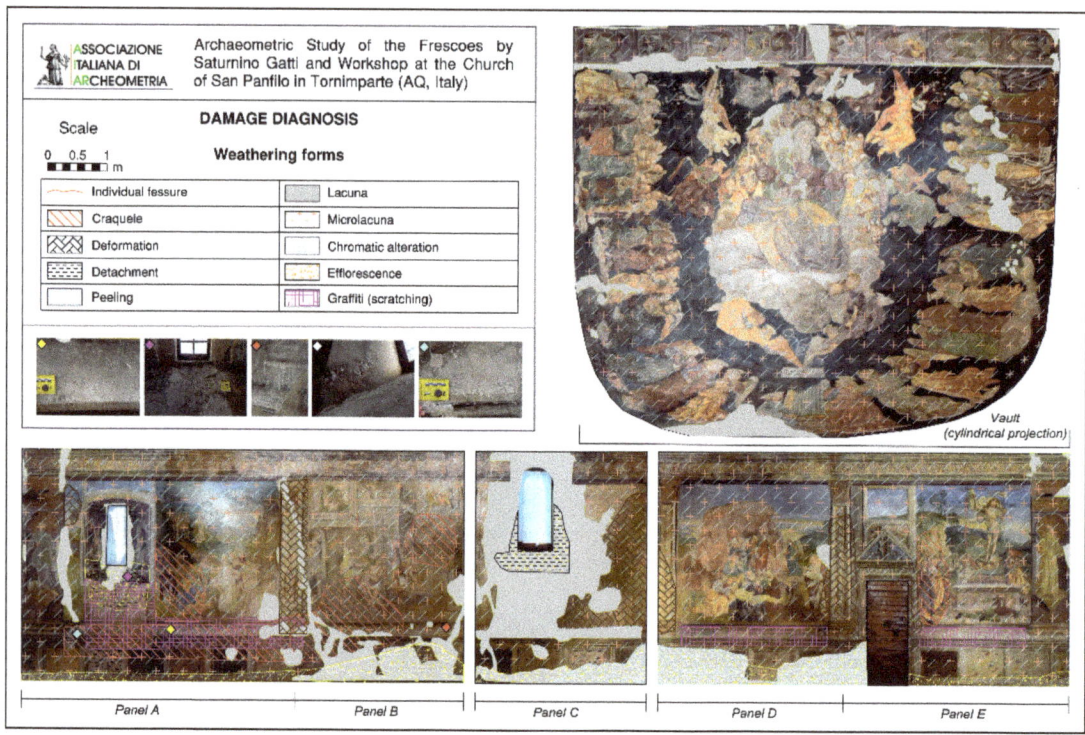

Figure 9. Distribution of the weathering forms affecting the five panels and the vault of the apsidal conch of St. Panfilo Church in Tornimparte (AQ, Italy). Coloured dots indicate the position of the acquired pictures in the wall paintings.

The walls of the apsidal conch show a very irregular morphology (surface warping), alternating concave and convex profiles (Figure S2). The deformation of these surfaces can be attributed both to the preparation of the plaster and to sub-efflorescence and/or rising damp, responsible for localized blistering. Sub-efflorescence and efflorescence, with consequent detachment and disintegration, can also be observed on the restoration mortars near the windows of panels A and C (Figure S2) and along the lower part of the apsidal conch. There the pictorial film applied as chromatic integration shows frequent peeling.

Figure 10 proposes a preliminary mapping of the damage categories identified taking into account the type of material (painted surface or restoration mortar) and the overlapping of the main weathering forms, giving more importance to those that are associated with loss and/or destabilization of material (cracks and deformations, detachments, *lacuna*, etc.). The damage index estimated for the whole examined surface is close to 3.2, while it assumes the highest values in panels A and B (DI_{lin} 3.6; DI_{prog} 3.7) and lower values in the vault (DI_{lin} 2.9; DI_{prog} 3.0). There are no particular differences between the linear and progressive indexes.

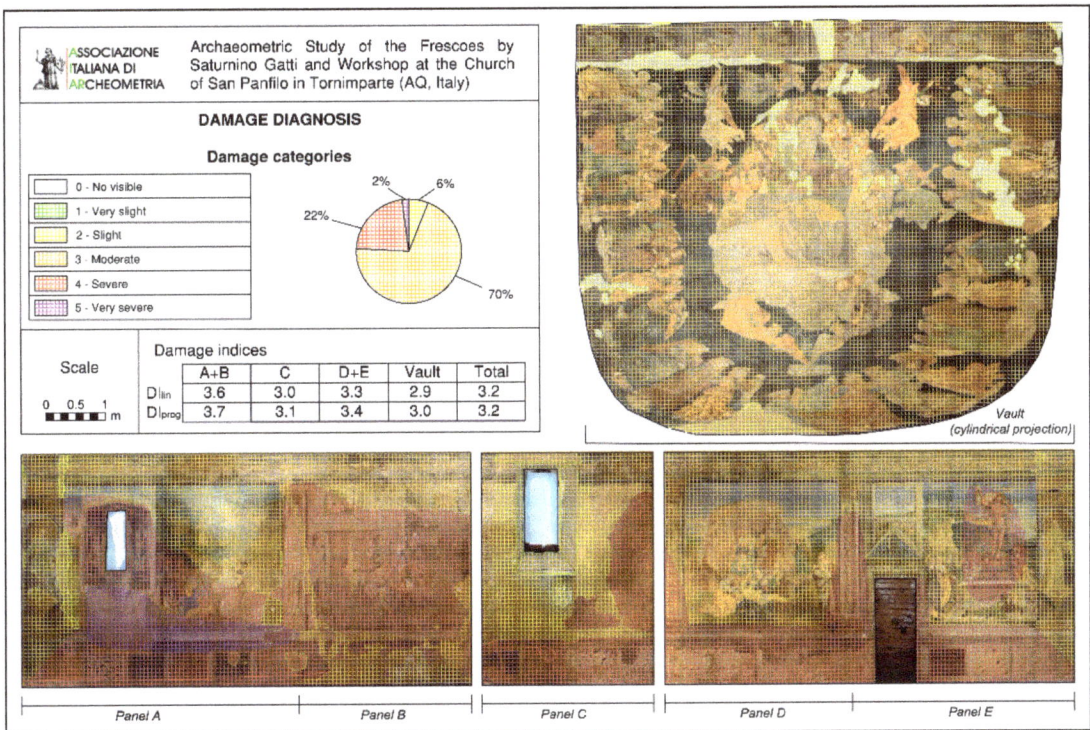

Figure 10. Distribution of the damage categories for the five panels and the vault of the apsidal conch of St. Panfilo Church at Tornimparte (AQ, Italy).

4. Discussion and Conclusions

Documentation of a monument is a fundamental step to correctly plan and perform any restoration interventions. For this reason, the 3D photogrammetry of the apsidal conch of St. Panfilo Church in Tornimparte (AQ, Italy) represented the first and essential survey to perform and made it possible to export a three-dimensional, navigable and explorable model, using an interactive file in pdf format available in this paper as supplementary material (Figure S1).

The same model is also visible on Sketchfab platform using the link https://sketchfab.com/models/88de7bbf64944408ad5fdda1d6eda141.

As a result of the 3D photogrammetric survey, it was possible to export 2D images (i.e., orthomosaics) and to import them into CAD software for diagnosis of damage to the five panels and the vault of the apsidal conch, providing maps of the distribution of materials, weathering forms and damage categories.

As far as the state of conservation of the painted surfaces is concerned, the decorated apsidal conch of St. Panfilo Church shows a significant overlap of damage pathologies probably attributable to both the executive techniques and the microclimate conditions of the site for which more detailed information can be found in the other contributions of this Special Issue [32,33,35,36]. In fact, the occurrence of "a secco" painting areas, generally considered a less durable executive technique over time, could be responsible for most of the detachments affecting the pictorial layer. The microclimate conditions (i.e., high relative humidity) could facilitate the rising damp and surface deposition of efflorescence or sub-efflorescence with localized blistering, detachment and disintegration phenomena.

The bad state of conservation of these paintings could also be due to previous invasive restorations that affected the surfaces with heavy retouching and repainting. It is especially observed close to the openings of panel A and panel C where modern mortar-based materials have been used, probably leading to the occurrence of the observed degradation phenomena [37–39].

The RAK photos of the examined panels were taken to document not only the decay phenomena of the mural painting surface but the painting technique and the preparatory drawings too. As a whole, the presence of incision marks (with minor pouncing), and the boundaries of *giornate* and *pontate*, clearly suggest the use of the *fresco* painting technique, although in several areas the color spread was *a secco*, especially where colour loss is observed. This fact is also suggested by the presence of a *morellone* layer under the blue sky background for the scene depicted on the panel E. Morellone was typically used in the *secco* painting technique. In this case, the blue pigment azurite (along with organic binder) was often used that easily turns into the green pigment malachite, as likely observed on panel A. In this area, thanks to the RAK photography, a clear example of rethinking depicting a monumental door and some religious buildings was observed.

It is also worthy of remark that the distribution of the *giornate* boundaries suggests that Saturnino Gatti generally had the tendency to separate the more detailed work of the figures of the scenes from the landscapes in the background. Nevertheless, a complete identification of the *giornate* boundaries was not possible since some of them were often hidden as a consequence of the several retouching and restoration works carried out over the centuries.

The present study shows how a documentation approach based on a multianalytical approach can provide a very important starting point for the restoration work of a complex monument, such as the wall painting in the church of St. Panfilo.

Lastly, as mentioned previously, this investigation represents a first digital documentation of the degradation phenomena affecting the Early Renaissance wall paintings realized by Saturnino Gatti during 1491–1494 in Tornimparte. The graphic contents here discussed, such as 3D photogrammetry (Figures 1 and S1), raking light photography (Figures 3 and 5–7) and damage mapping (Figures 8–10) are able to illustrate the state of conservation of the monument before its restoration.

Supplementary Materials: The following supporting information can be downloaded at: https://www.mdpi.com/article/10.3390/app13095689/s1, Figure S1: interactive file in pdf format; Figure S2: weathering forms affecting the wall paintings.

Author Contributions: Conceptualization, data curation, writing—original draft preparation. F.B., S.C., F.I., L.L., C.P. and A.P.S.; Investigation, visualization, F.B., S.C., C.G., F.I., L.L., S.P., C.P. and A.P.S.; Supervision F.I., A.P.S. and C.P. All authors have read and agreed to the published version of the manuscript.

Funding: This work was performed in the framework of the Project Tornimparte—"Archaeometric investigation of the pictorial cycle of Saturnino Gatti in Tornimparte (AQ, Italy)" sponsored in 2021 by the Italian Association of Archaeometry AIAR (www.associazioneaiar.com).

Institutional Review Board Statement: Not applicable.

Informed Consent Statement: Not applicable.

Data Availability Statement: The data presented in this study are available on request from the authors.

Acknowledgments: The authors thank Domenico Fusari, President of PRO LOCO Association of Tornimparte, and Giacomo Carnicelli, Mayor of the municipality of Tornimparte for their support during the work activities. Moreover, the authors wish to give special thanks to the conservator Maria Fernanda Falcon Martinez and to the art historian Saverio Ricci of the *Soprintendenza Archeologica, Belle Arti e Paesaggio per le province di l'Aquila e Teramo* for their valuable suggestions and help during the measurements and in situ acquisitions.

Conflicts of Interest: The authors declare no conflict of interest.

References

1. Arbace, L.; Di Paolo, G. *I Volti Dell'anima, Saturnino Gatti: Vita e Opere Di Un Artista Del Rinascimento*; Paolo De Siena Editore: Pescara, Italy, 2012; ISBN 8896341116.
2. Mannetti, T.R.; Chelli, N.; Vecchioli, G. *Saturnino Gatti Nella Chiesa Di San Panfilo a Tornimparte*; L'Aquila Publishing: L'Aquila, Italy, 1992.
3. Soler, F.; Melero, F.J.; Luzón, M.V. A Complete 3D Information System for Cultural Heritage Documentation. *J. Cult. Herit.* **2017**, *23*, 49–57. [CrossRef]
4. Guarnieri, A.; Pirotti, F.; Vettore, A. Cultural Heritage Interactive 3D Models on the Web: An Approach Using Open Source and Free Software. *J. Cult. Herit.* **2010**, *11*, 350–353. [CrossRef]
5. Yastikli, N. Documentation of Cultural Heritage Using Digital Photogrammetry and Laser Scanning. *J. Cult. Herit.* **2007**, *8*, 423–427. [CrossRef]
6. Yilmaz, H.M.; Yakar, M.; Gulec, S.A.; Dulgerler, O.N. Importance of Digital Close-Range Photogrammetry in Documentation of Cultural Heritage. *J. Cult. Herit.* **2007**, *8*, 428–433. [CrossRef]
7. Apollonio, F.I.; Basilissi, V.; Callieri, M.; Dellepiane, M.; Gaiani, M.; Ponchio, F.; Rizzo, F.; Rubino, A.R.; Scopigno, R.; Sobra', G. A 3D-Centered Information System for the Documentation of a Complex Restoration Intervention. *J. Cult. Herit.* **2018**, *29*, 89–99. [CrossRef]
8. Pelosi, C.; Calienno, L.; Fodaro, D.; Borrelli, E.; Rubino, A.R.; Sforzini, L.; Monaco, A. Lo An Integrated Approach to the Conservation of a Wooden Sculpture Representing Saint Joseph by the Workshop of Ignaz Günther (1727–1775): Analysis, Laser Cleaning and 3D Documentation. *J. Cult. Herit.* **2016**, *17*, 114–122. [CrossRef]
9. Means, B.K. 3D Recording, Documentation and Management of Cultural Heritage. *Hist. Archaeol.* **2017**, *51*, 582–583. [CrossRef]
10. UNI-EN16095; Conservation of Cultural Property: Condition Recording for Movable Cultural Heritage. European Committee for Standardization: Brussels, Belgium, 2012.
11. Bentkowska-Kafel, A.; MacDonald, L. *Digital Techniques for Documenting and Preserving Cultural Heritage*; Arc Humanities Press: Yorkshire, UK, 2017; ISBN 9781942401346.
12. Pavlidis, G.; Koutsoudis, A.; Arnaoutoglou, F.; Tsioukas, V.; Chamzas, C. Methods for 3D Digitization of Cultural Heritage. *J. Cult. Herit.* **2007**, *8*, 93–98. [CrossRef]
13. Agosto, E.; Bornaz, L. 3D Models in Cultural Heritage: Approaches for Their Creation and Use. *Int. J. Comput. Methods Herit. Sci.* **2017**, *1*, 1–9. [CrossRef]
14. Xiao, W.; Mills, J.; Guidi, G.; Rodríguez-Gonzálvez, P.; Gonizzi Barsanti, S.; González-Aguilera, D. Geoinformatics for the Conservation and Promotion of Cultural Heritage in Support of the UN Sustainable Development Goals. *ISPRS J. Photogramm. Remote Sens.* **2018**, *142*, 389–406. [CrossRef]
15. Cosentino, A.; Gil, M.; Ribeiro, M.; Di Mauro, R. Technical Photography for Mural Paintings: The Newly Discovered Frescoes in Aci Sant'Antonio (Sicily, Italy). *Conserv. Património* **2014**, *20*, 23–33. [CrossRef]
16. Zhang, J.; Kang, K.; Liu, D.; Yuan, Y.; Yanli, E. Vis4Heritage: Visual Analytics Approach on Grotto Wall Painting Degradations. *IEEE Trans. Vis. Comput. Graph.* **2013**, *19*, 1982–1991. [CrossRef] [PubMed]
17. Chai, C.; de Brito, J.; Gaspar, L.; Silva, P.A. Predicting the Service Life of Exterior Wall Painting: Techno-Economic Analysis of Alternative Maintenance Strategies. *J. Constr. Eng. Manag.* **2014**, *140*, 4013057. [CrossRef]
18. Langella, A.; Calcaterra, D.; Cappelletti, P.; Ciarcia, S.; D'Amore, M.; Di Martire, D.; Graziano, S.F.; de Gennaro, M. An Example of Integrated Geological Survey of Geomaterials and Their Weathering Forms: The Reggia Di Caserta Main Façade. *Stud. Conserv.* **2022**, 1–13. [CrossRef]
19. Fitzner, B. Documentation and Evaluation of Stone Damage on Monuments. In Proceedings of the 10th International Congress on Deterioration and Conservation of Stone, Stockholm, Sweden, 27 June–2 July 2004; Volume 27, pp. 677–690.
20. Fitzner, B.; Heinrichs, K. Damage Diagnosis at Stone Monuments—Weathering Forms, Damage Categories and Damage Indices. *Acta Univ. Carol. Geol.* **2001**, *1*, 1–49.
21. Fitzner, B.; Heinrichs, K.; La Bouchardiere, D. Damage Index for Stone Monuments. In Proceedings of the Protection and Conservation of the Cultural Heritage of the Mediterranean Cities. In Proceedings of the 5th International Symposium on the Conservation of Monuments in the Mediterranean Basin, Sevilla, Spain, 5–8 April 2000.
22. Grifa, C.; Barba, S.; Fiorillo, F.; Germinario, C.; Izzo, F.; Mercurio, M.; Musmeci, D.; Potrandolfo, A.; Santoriello, A.; Toro, P.; et al. The Domus of Octavius Quartio in Pompeii: Damage Diagnosis of the Masonries and Frescoed Surfaces. *Int. J. Conserv. Sci.* **2016**, *7*, 885–900.
23. Izzo, F.; Furno, A.; Cilenti, F.; Germinario, C.; Gorrasi, M.; Mercurio, M.; Langella, A.; Grifa, C. The Domus Domini Imperatoris Apicii Built by Frederick II along the Ancient Via Appia (Southern Italy): An Example of Damage Diagnosis for a Medieval Monument in Rural Environment. *Constr. Build. Mater.* **2020**, *259*, 119718. [CrossRef]
24. Germinario, C.; Gorrasi, M.; Izzo, F.; Langella, A.; Limongiello, M.; Mercurio, M.; Musmeci, D.; Santoriello, A.; Grifa, C. Damage Diagnosis of Ponte Rotto, a Roman Bridge along the Ancient Appia. *Int. J. Conserv. Sci.* **2020**, *11*, 277–290.
25. Randazzo, L.; Collina, M.; Ricca, M.; Barbieri, L.; Bruno, F.; Arcudi, A.; La Russa, M.F. Damage Indices and Photogrammetry for Decay Assessment of Stone-Built Cultural Heritage: The Case Study of the San Domenico Church Main Entrance Portal (South Calabria, Italy). *Sustainability* **2020**, *12*, 5198. [CrossRef]

26. Galli, A.; Alberghina, M.F.; Re, A.; Magrini, D.; Grifa, C.; Ponterio, R.C.; La Russa, M.F. Special Issue: Results of the II National Research Project of AIAr: Archaeometric Study of the Frescoes by Saturnino Gatti and Workshop at the Church of San Panfilo in Tornimparte (AQ, Italy). *Appl. Sci.* 2023; *to be submitted*.
27. Lowe, D.G. Distinctive Image Features from Scale-Invariant Keypoints. *Int. J. Comput. Vis.* **2004**, *60*, 91–110. [CrossRef]
28. Lanteri, L.; Agresti, G.; Pelosi, C. A New Practical Approach for 3D Documentation in Ultraviolet Fluorescence and Infrared Reflectography of Polychromatic Sculptures as Fundamental Step in Restoration. *Heritage* **2019**, *2*, 207–215. [CrossRef]
29. Garcia de Miguel, J.M. ICOMOS Illustrated Glossary on Stone Deterioration Patterns Glosario Ilustrado de Formas de Deterioro de La Piedra. In *Monuments & Sites*; ICOMOS: Paris, France, 2011; Volume 15.
30. CNR; ICR. *Raccomandazioni Normal: 1/88 Alterazioni Macroscopiche Dei Materiali Lapidei: Lessico*; CNR; ICR: Rome, Italy, 1991.
31. Colantonio, C.; Lanteri, L.; Ciccola, A.; Serafini, I.; Postorino, P.; Censorii, E.; Rotari, D.; Pelosi, C. Imaging Diagnostics Coupled with Non-Invasive and Micro-Invasive Analyses for the Restoration of Ethnographic Artifacts from French Polynesia. *Heritage* **2022**, *5*, 215–232. [CrossRef]
32. Germinario, L.; Giannossa, L.C.; Lezzerini, M.; Mangone, A.; Mazzoli, C.; Pagnotta, S.; Spampinato, M.; Zoleo, A.; Eramo, G. Petrographic and Chemical Characterization of the Frescoes by Saturnino Gatti (Central Italy, 15th Century): Microstratigraphic Analyses on Thin Sections. *Appl. Sci.* 2023; *to be submitted*.
33. Calandra, S.; Centauro, I.; Laureti, S.; Ricci, M.; Salvatici, T.; Sfarra, S. Sonic, Humidity and Thermal Imaging Methods for the Analysis of Decaying Frescoes of Apsidal Conch of San Panfilo Church. *Appl. Sci.* 2023; *to be submitted*.
34. Briani, F.; Caridi, F.; Ferella, F.; Gueli, A.M.; Marchegiani, F.; Nisi, S.; Paladini, G.; Pecchioni, E.; Politi, G.; Santo, A.P.; et al. Multi-Technique Characterization of Painting Drawings of the Pictorial Cycle at the San Panfilo Church in Tornimparte. *Appl. Sci.* 2023; *to be submitted*.
35. Comite, V.; Bergomi, A.; Lombardi, C.A.; Fermo, P. Characterization of Soluble Salts on the Frescoes by Saturnino Gatti in the Church of San Panfilo in Villagrande Di Tornimparte (L'Aquila). *Appl. Sci.* 2023; *to be submitted*.
36. Ferrarese, S.; Bertoni, D.; Golzio, A.; Lanteri, L.; Pelosi, C.; Re, A. Microclimate Analysis of the San Panfilo Church in Tornimparte, Italy. *Appl. Sci.* 2023; *to be submitted*.
37. Cultrone, G.; Arizzi, A.; Sebastián, E.; Rodriguez-Navarro, C. Sulfation of Calcitic and Dolomitic Lime Mortars in the Presence of Diesel Particulate Matter. *Environ. Geol.* **2008**, *56*, 741–752. [CrossRef]
38. Izzo, F.; Grifa, C.; Germinario, C.; Mercurio, M.; De Bonis, A.; Tomay, L.; Langella, A. Production Technology of Mortar-Based Building Materials from the Arch of Trajan and the Roman Theatre in Benevento, Italy. *Eur. Phys. J. Plus* **2018**, *133*, 363. [CrossRef]
39. Lubritto, C.; Ricci, P.; Germinario, C.; Izzo, F.; Mercurio, M.; Langella, A.; Cuenca, V.S.; Torres, I.M.; Fedi, M.; Grifa, C. Radiocarbon Dating of Mortars: Contamination Effects and Sample Characterisation. The Case-Study of Andalusian Medieval Castles (Jaén, Spain). *Meas. J. Int. Meas. Confed.* **2018**, *118*, 362–371. [CrossRef]

Disclaimer/Publisher's Note: The statements, opinions and data contained in all publications are solely those of the individual author(s) and contributor(s) and not of MDPI and/or the editor(s). MDPI and/or the editor(s) disclaim responsibility for any injury to people or property resulting from any ideas, methods, instructions or products referred to in the content.

Article

Identifying Original and Restoration Materials through Spectroscopic Analyses on Saturnino Gatti Mural Paintings: How Far a Noninvasive Approach Can Go

Letizia Bonizzoni [1,*], Simone Caglio [2], Anna Galli [2], Chiara Germinario [3], Francesco Izzo [4,5] and Donata Magrini [6]

1. Dipartimento di Fisica Aldo Pontremoli, Università degli Studi di Milano, Via Celoria 16, 20133 Milano, Italy
2. Dipartimento di Scienza dei Materiali, Università degli Studi di Milano-Bicocca, 20125 Milano, Italy; simone.caglio@unimib.it (S.C.); anna.galli@unimib.it (A.G.)
3. Department of Science and Technology, University of Sannio, Via de Sanctis, 82100 Benevento, Italy; chiara.germinario@unisannio.it
4. Department of Earth Sciences, Environment and Resources, Federico II University, Via Cinthia, 80126 Naples, Italy; francesco.izzo4@unina.it
5. CRACS—Center for Research on Archaeometry and Conservation Science, 80126 Naples, Italy
6. ISPC-CNR, Istituto di Scienze del Patrimonio Culturale, CNR, Area della Ricerca di Firenze, Via Madonna del Piano, 10, 50019 Sesto Fiorentino, Italy; donata.magrini@cnr.it
* Correspondence: letizia.bonizzoni@unimi.it

Abstract: This paper presents the results obtained for the mural paintings (XV century CE) in the church of San Panfilo in Villagrande di Tornimparte (AQ, Italy) by means of noninvasive spectroscopic techniques; this research is a part of the project on the Saturnino Gatti pictorial cycle, promoted and coordinated by the AIAr (the Italian Archaeometry Association). Digital optical microscopy (OM), X-ray fluorescence spectroscopy (XRF), fiber optics reflectance spectroscopy in the UV–Vis–NIR range (FORS), Fourier transform infrared spectroscopy in the external reflection mode (ER-FTIR), and Raman spectroscopy were performed on the points selected based on the image analysis results and the few available records on previous intervention, with the aim of characterizing both the original and restoration organic and inorganic materials. The synergic application of complementary techniques allowed us to obtain a complete picture of the palette and the main alteration products and organic substances (of rather ubiquitous lipid materials and less widespread resin and proteinaceous materials in specific points). The identification of modern compounds permitted the individuation of restoration areas; this was confirmed by the comparison with multiband imaging results, as in the case of specific green and blue pigments, strictly related to the presence of high signals of zinc. This analytical protocol left only very few ambiguities and allowed to minimizing the number of samples taken to clarifying, by sample laboratory analyses, the few doubts still open.

Keywords: mural painting; Saturnino Gatti; Italian Archaeometry Association; XRF; FORS; ER-FTIR; Raman spectroscopy; pigments; binders

1. Introduction

This paper contributes to the Special Issue, Results of the II National Research Project of AIAr: Archaeometric Study of the Frescoes by Saturnino Gatti and Workshop at the Church of San Panfilo in Tornimparte (AQ, Italy), in which the scientific results of the II National Research Project, conducted by members of the Italian Association of Archaeometry (AIAr), are discussed and collected. For in-depth details on the project's aims, see the introduction of the Special Issue [1].

It is worth explaining that Tornimparte, its hamlet Villagrande, and the entire L'Aquila region suffered a major earthquake in 1703. We do not have a precise record of the damage suffered by the buildings, but it is quite relevant to the fact that the present houses are all post-1700, and the churches are all designed in the Baroque style. Indeed, San Panfilo in

Villagrande is a remarkable exception; however, we can speculate it was damaged by the disaster, too. This same region was affected by another great earthquake in 1915 and then again in 2009, which devastated the whole area.

The ultimate goal of the project is to support the foreseen restoration interventions following the last earthquake through the complete study of the conservation status and the characterization of the original materials, the superimposed ones, and the degradation phenomena affecting the wall paintings. The project involved several Italian groups working on cultural heritage materials: for each aspect of the research, a selected team of researchers from the participant groups was involved. The analyses performed within the project comprise micro-climatic monitoring, multispectral imaging and photogrammetric surveying, degradation mapping, and thermography investigations, alongside the in situ and laboratory material characterizations.

One of the first steps of the study's project of the pictorial cycle, attributed to Saturnino Gatti at the church of San Panfilo in Villagrande di Tornimparte, is the one presented in this paper. Indeed, immediately after the imaging analyses (used as exploratory techniques for the choice of the region of interest for our study and as the first approach to distinguish between the original and restored areas), noninvasive spectroscopic techniques were applied with the aim of characterizing both original and restoration organic and inorganic paint materials. Namely, digital optical microscopy (OM), X-ray fluorescence spectroscopy (XRF), fiber optics reflectance spectroscopy in the UV–Vis–NIR range (FORS), Fourier transform infrared spectroscopy in the external reflection mode (ER-FTIR), and Raman spectroscopy were used. Each one of these techniques has its own place in the cultural heritage field, and they are often used together in order to overcome specific limits. The synergic application of analytical techniques, exploiting different radiation frequency ranges, indeed allows for obtaining complementary information for a complete picture of the palette [2–5], main alteration products [6–9] and organic substances [10–14]. Despite the poor conservation state of the decorative materials and the presence of several retouched and restored areas, the analytical approach adopted here provided very informative results, not only for the identification of original materials but also for the identification of modern compounds, allowing the individuation of restored areas that can be confirmed and highlighted by results obtained by imaging results [15]. In the present contribution, preliminary outputs are illustrated, and allowed to suggest a few areas, then selected for additional sampling and laboratory analyses.

2. Materials and Methods

Tornimparte Church, the parish church located in Villagrande, a hamlet of Tornimparte, is dedicated to Panfilo di Sulmona and dates back to around the year 1000. It still retains the original architectural forms, although profound renovations were made following the 1461 and the 1703 L'Aquila earthquakes, being a building located in an area with a high seismic risk. The most important intervention, however, took place in 1495, when Saturnino Gatti, a local young artist that can be fully considered an exponent of the Renaissance, painted in the aspis of the church a sequence of frescoes. The artistic work in Tornimparte Church has miraculously survived until today and is essentially the only testimony of Saturnino Gatti's mural paintings, while several of his easel paintings and sculptures still survive.

The frescoes underwent a number of restoration interventions: the first one we know about dates back to 1929, and even if there are no written documents describing it, the existence of still pictures allows us to understand that restorers worked mainly on the faces of several characters [16]. It is worth noting that these same areas are nowadays very damaged. A second known intervention was then made in 1951, and the last one in 1972, following a fire caused by a short circuit in 1958. Starting from this problematic condition, the evaluation of material characterization comes to light as a necessary step before further restoration works following the 2009 earthquake and needs to be in cooperation with the local government authorities for cultural heritage (Segretariato regionale del ministero per i

Beni e le Attività Culturali e per il Turismo per l'Abruzzo, and Soprintendenza Archeologia, Belle Arti e Paesaggio per la Città dell'Aquila e i comuni del Cratere).

2.1. Selected Areas

The mural paintings of San Panfilo Church were created around 1495 by Saturnino Gatti. In the middle of the vault is the representation of Paradise: the Eternal Father is the central figure surrounded by Angels and the Blessed. In the arch above the main altar, the Prophets who predicted the coming of the Redeemer are depicted, while on the side, Archangel Gabriel is shown in the act of announcing the birth of the Son of God to the Virgin. Around the apsis, in five separate panels, the moments of the Redemption are reproduced: the capture of Jesus and the kiss of Judas in the Garden of Olives, the scourging, the crucifixion, the Deposition of Christ, and the Resurrection of the Savior.

The measuring points were selected on three of the above-described panels of the lower part of the apsis and on the blue background on the vault. For a more detailed description of the mural paintings, please refer to [1]. The panels considered for the noninvasive spectroscopic investigations were the Garden of Olives (panel A in the following), the Deposition of Christ (panel E in the following), and the Resurrection (panel D in the following). The selection of measuring points was made with the supervision and support of the Superintendent Inspectors of the Italian Ministry of Culture and the restorers, taking into account the geometrical limitations for the movements of the instruments. The selected regions are representative of the original painting layers and of the various restoration interventions, either documented or supposed; the panels and the investigated points (about 60 for the entire cycle) are reported in Figure 1. Moreover, thanks to the close coordination of the project, this noninvasive approach was performed after and on the basis of the results obtained by imaging techniques [17]; in particular, visible images under ultraviolet light and infrared reflectography were used to serve as exploratory techniques, allowing to recognize regions of interest for the study and to distinguish between the original and restored sections in an artwork.

Figure 1. *Cont.*

Figure 1. Pictures of the analyzed scenes and investigated points: (**a**) panels on the vault; (**b**–**d**) analyzed points on the vault; (**e**) the Resurrection, details of investigated points; (**f**) Garden of Olives, investigated points; (**g**,**h**) the Resurrection, details and investigated points.

2.2. Analytical Techniques

As already stated, several complementary noninvasive techniques were applied, as detailed in the following paragraphs. For obvious reasons, the exploitation of portable instrumentations and techniques is highly preferable: the ultimate goal of the present campaign was to acquire as much information as possible about both original and restoring organic and inorganic materials in order to minimize the number of samples required for laboratory investigations.

2.2.1. Optical Digital Microscopy (OM) in Visible Light

Magnified images were obtained by a Dino-Lite portable digital microscope, with a CCD sensor of 5 Mpx for each point of analysis. The 50× and 200× magnifications were considered. For a correct focus on the surface, the microscope head was gently placed on the wall.

2.2.2. Fiber Optics Reflectance Spectroscopy in the UV–Vis–NIR Range (FORS)

Fiber optics reflectance spectroscopy in the UV–Vis–NIR range was performed using a FORS spectrometer (StellarNet BlueWave) working in the 400–1000 nm spectral range, with a resolution of 0.5 nm. The light source was a tungsten halogen lamp (2800 K, 200 W/m^2 power). The instrument allows the measurements in different geometry configurations and through different configurations of fiber optics. Due to the rough and opaque surface of the wall painting, to maximize the signal to be acquired, a 0/0 geometry (perpendicular to the wall surface and gently applied to the investigated areas) was chosen, exploiting a bifurcated fiber optic, which allows bringing the light to the sample and collecting the response signal from the same probe. To reduce the signal-to-noise ratio, an integration time of 500 milliseconds was used with the average over three scans.

2.2.3. Raman Spectroscopy

Raman spectra were acquired by means of an i-Raman Plus BW Tec in the fiber optics configuration; the fiber optics head, mounted in a plastic holder, was gently placed on the wall. The mounted laser had a wavelength of 785 nm with 320 mW of maximum power; the spectral range was 200–3000 cm^{-1} with a resolution lower than 5 cm^{-1}. The laser power used was 8%, with a 10 s integration time and 3 accumulations for all the colors except the red ones, for which the power was 10% with a 10 s integration time and 5 accumulations.

2.2.4. Fourier Transform Infrared Spectroscopy

Infrared spectra were obtained via Fourier transform infrared spectroscopy (FTIR) on panels A, D, and E by using a Bruker Alpha portable FTIR spectrometer with an external reflectance (ER) module and equipped with a ROCKSOLID™ interferometer and a ZnSe/KBr beam splitter with a DTGS detector (room temperature). Circular areas of about 3 mm in diameter were analyzed by gently leaning the ER module on the wall for non-destructive analyses. The spectra were collected in the spectral range between 7500 and 400 cm^{-1}, with a resolution of 4 cm^{-1}, and a number of scans, variable from 64 to 192 (1–3 min). Opus 7.2 software was used for data acquisition and processing, treated by smoothing and log transformation (A' = log (1/R) [18].

2.2.5. X-ray Fluorescence Spectroscopy

Two different spectrometers were used for the XRF in situ analyses: one for the paintings on the walls of the apsis, and the other was a handheld model, which allowed working on the scaffolding for the vault.

The elementary analysis of the apsis walls was performed using a portable XRF Assing LITHOS 3000 spectrometer equipped with a Mo anode X-ray tube and a Peltier-cooled Si-PIN detector. The radiation is quasi-monochromatic at the energy of Mo Kα energy by means of a Zr transmission filter; the X-ray tube mounts a 4 mm diameter collimator so that the irradiated area on the sample is about 25 mm^2. Data were collected using 25 kV high voltage and 0.3 mA tube current, with a 100 s acquisition time. The distance between the head of the device and the wall was about 10 mm.

The XRF analyses on the vault were instead performed using a Tracer III SD Bruker portable device equipped with a rhodium X-ray tube, a palladium anticathode, and a solid-state silicon detector energy dispersion system. The set-up was as follows: 40 keV and 12 μA for 60 s; the instrument head, covered in soft plastic, was gently applied on the vault. The measuring area was an elliptical spot of 4 mm × 7 mm. For data visualization and fitting, ARTAX software was used.

For both spectrometers, we based our considerations on the raw spectra fitting: we assigned each characteristic energy to the related chemical element and performed the relative intensity evaluation to acquire information about the chemical composition of the materials. From a methodological point of view, a comparison between the results obtained by two different commercial portable XRF spectrometers mounting X-ray tubes with different anode materials allows us to obtain the most from this technique. Indeed, experience suggests that spectrometers should be selected according to the specific requirements and type of analysis to perform, always keeping in mind the difficulty of also working on hard-to-access surfaces, such as the Tormiparte vault frescoes for which scaffolds were required [19].

3. Results and Discussion

The goal of the research was to provide as much information as possible by exploiting a noninvasive approach. It is well known that studying a mural painting is challenging for many reasons, the most important of which is the contemporary presence of plaster, pigments, degradation products, pollution, and restorations. Analyses virtually carry a huge amount of information, which is difficult to disentangle, as it is not trivial to clearly link an element or a compound to the corresponding material.

The in situ spectroscopic analyses have to be considered a necessary but insufficient condition for a complete characterization of the pictorial materials and techniques applied. They certainly have the advantage of helping in the mapping of pigment and binder presence. In addition, the joint use of the various spectroscopic techniques partially compensates for the intrinsic impossibility of uniquely identifying the present stratigraphy provided by the analyses performed on the micro-samples.

We thus performed the already mentioned analytical point-wise spectroscopic investigations and then compared the results with imaging investigations [17]. We thus inferred the type of materials (pigments and binders) and were able to distinguish between the original and restored areas. The association of the selected techniques makes it possible to overcome the limitations of each one and to detect all classes of pigments, binders, and alteration products. By using these results, areas for the samplings were then chosen to remove all remaining doubts.

3.1. Plaster

Spectra, acquired via ER-FTIR, detected the ubiquitous presence of calcium carbonate, the principal constituent of the wall paintings' support, featured by the undistorted $\nu_1 + \nu_3$ combination band of the $(CO_3)^{2-}$ group at ca. 2510 cm^{-1} (with a shoulder at ca. 2590 cm^{-1}), and the *Reststrahlen* and derivative effects at ca. 1410 cm^{-1} (with an asymmetric stretching band of CO_3), 873 cm^{-1} (with an "out-of-plane" bending vibration of the $(CO_3)^{2-}$ group), and 713 cm^{-1} (with an "in-plane" bending vibration of the $(CO_3)^{2-}$ group) [20]. Calcium and calcium carbonate in white areas and several color backgrounds were also detected, respectively, by XRF and Raman spectroscopy, as expected.

3.2. Pigments

The investigated areas were chosen together with the art historians and conservators examining the results of previously performed imaging analyses—both to have a complete picture of the original and added materials and to answer specific questions. In general, typical pigments of mural paintings were detected. Indeed, yellow and brownish hues were obtained by iron oxides and/or hydroxides, as well as red pigments, which constituted mostly of iron oxides. Only in a few cases were the red details executed by using vermillion, as, for instance, the darker vests seen in the A and E panels [1]. This pigment is easily recognized based on Raman spectra and from the presence of mercury in the XRF ones. In the left box, in Figure 2, the digital optical microscope image of a vest painted with vermillion in panel A is reported.

Flesh tones were obtained by mixing a calcium-based white pigment (see Section 3.2.2 for white characterization) with iron oxides and vermillion traces. In the restored areas, highlighted by the presence of high zinc signals in the XRF spectra and confirmed by imaging investigations [17], lead white was also detected based on the presence of high lead signals in the XRF spectra. The digital optical microscope image of a restored flesh tone in panel A is reported in the right box in Figure 2.

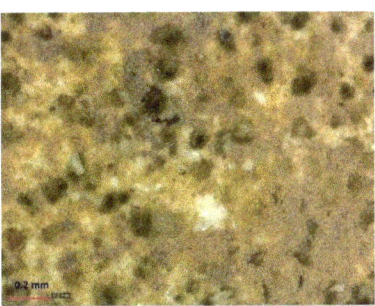

Figure 2. Digital optical microscope images: on the (**left**), panel A, dark red vest (ochre with vermillion); on the (**right**), panel A, flesh tone, restored area (ochres, lead white, and zinc white). Red scale bars correspond to 0.2 mm.

In the following sections, a deeper overview of the most interesting cases is provided. All the results are summarized in Table 1.

Table 1. Summary table of the pigment detected in the different investigated areas on panels A, D, and E.

Panel	Color	Pigments
Panel A, E	Green	Green earth with small quantities of a copper-based pigment
Panel A	Green	Green earth with zinc white with small quantities of chromium-based pigment
Panel A	Green	Green earth, zinc, and lead white, chromium-based pigment with small quantities of copper-based pigment
Panel E	Green	Green earth and other iron oxides
Panel A	Green	Copper-based pigment with lead white
Panel D	Green	Copper-based pigment, green earth, and zinc white
Panel E	Green	Copper-based pigment with green earth and zinc white
Panel A, D, E	Flesh tones	Iron oxides
Panel A	Flesh tones	Iron oxides with zinc and lead white
Panel A, E	Blue	Azurite
Panel A	Blue	Ultramarine with zinc and lead white
Panel D	Blue	Ultramarine with zinc white
Panel A	Blue	Prussian blue with zinc white
Panel A, D	Red	Iron oxides
Panel A, E	Red	Iron oxides with small quantities of vermillion
Panel E	Red	Vermillion
Panel A, D, E	Yellow	Yellow ochre
Panel D	Yellow	Yellow ochre and zinc white
Panel E	Yellow	Yellow ochre with small quantities of vermillion
Panel E	Brown	Iron oxides
Panel A, D, E	White	Calcite
Panel A	White	Calcite with small quantities of zinc white
Panel D	Black	Carbon or bone black

3.2.1. Green

At least two different past restorations are evident on the bases of the detected green pigments, where one was characterized by the use of a copper-based pigment and the other by a chromium-based green. Indeed, copper-based green and green earth were detected, either coupled in the same measured point or separated in different points (see Table 1 for details). This could possibly be linked to various restoration phases and can be related to imaging mapping [17]. For instance, in panel D, both copper-based green and green earth (this last supposed was to be the original material for greens) were detected together with high zinc signals in the XRF spectra, indicating a modern intervention, while in one detail on panel A, only copper-based green was found together with high signals of zinc, indicating this pigment was surely used for restoration intervention. On the other hand, in the restored green areas of panel A, characterized by a high zinc presence, significant chromium signals were detected through XRF analysis. From the in situ noninvasive investigations, the green points left several open questions; for this reason, the green areas were chosen for sampling and laboratory analyses [21].

3.2.2. White

In all the investigated white areas, the analytical techniques applied indicate the presence of calcite, detected both as a calcium presence by XRF and as calcium carbonate by Raman spectroscopy. It is worth stressing that lead white was never detected in the white areas of the analyzed panels; however, it was sometimes detected (see Table 1) in the restored areas of the flesh tones (see Figure 2 right) coupled with the presence of zinc and thus related to modern restorations. Indeed, the presence of zinc white was mapped by UV fluorescence [17] as it exhibits a well-visible lemon-yellow fluorescence. The obtained images indicate a massive use of this pigment for retouching painting.

In the white shades, and in particular, in the highlights, the spectral evidence of proteinaceous materials was unveiled by ER-FTIR (Figure 3). Proteins were recognized by the stair-step pattern of the diagnostic absorption bands (Figure 3a) due to the stretching vibration of the C=O bond in the carbonyl group at ca. 1635 cm^{-1} (amide I), affected by the derivative distortion in external reflection, and the C-N stretching and N-H bending vibrations at ca. 1530 cm^{-1} (amide II). Moreover, they show the asymmetric stretching vibrations of methyl CH_3-CH_2 groups, respectively, at ca. 2980, 2930, and 2870 cm^{-1} [22,23].

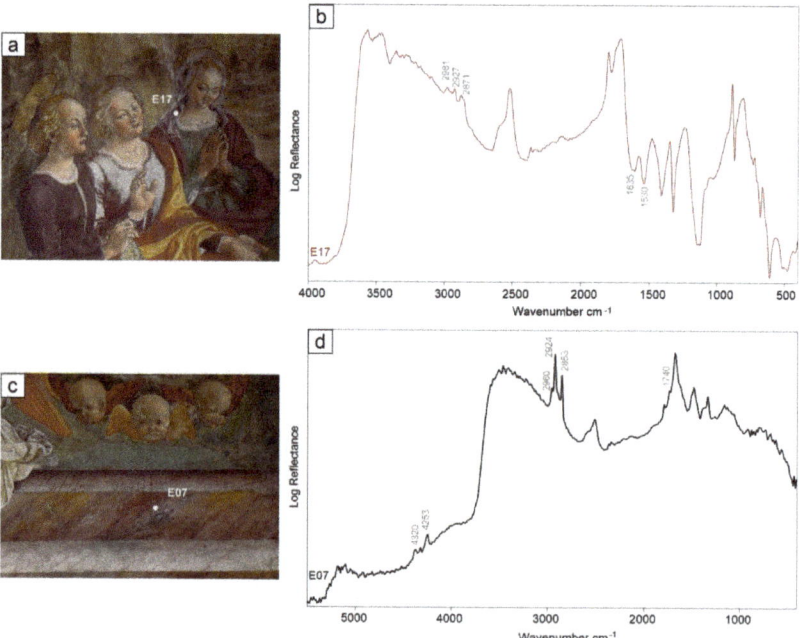

Figure 3. Infrared spectra, acquired at different points of panel E, reveal the use of proteinaceous materials for the highlights (**a**,**b**) and lipids for the gildings (**c**,**d**).

3.2.3. Blue

The combination of XRF, FORS, and Raman spectroscopy allowed the detection of several blue pigments related to the original and restoration areas pertaining to various interventions, according to the imaging analysis results [17]. On the panels, both azurite and ultramarine were, in fact, detected in different areas, as highlighted by the combined use of the spectroscopic techniques and reported in Table 1. In particular, XRF can detect high copper contents in azurite, while FORS can clearly distinguish the two different materials. It is interesting to note that, both in panel A and in panel D, the presence of ultramarine is related to a high presence of zinc, as indicated by the XRF spectra. It is worth noting that azurite was mapped using the UV fluorescence images [17], thanks to the binder, as this pigment was usually applied using the "secco" technique [24,25]. On panels A and E, the typical infrared bands of azurite, at ca. 2592, 2557, and 2510 cm^{-1} ($v_1 + v_3$ combination band of $(CO_3)^{2-}$), along with the strong doublet at ca. 4380 and 4244 cm^{-1} (combination $v + \delta$ (OH) and overtone $3v_3$), were detected by ER-FTIR (Figure 4) [26]. In these areas, the XRF (high copper presence) and FORS spectroscopy indicated the use of this specific blue pigment. For blue pigments, as well as for green ones, a second restoration phase was highlighted by the use of Prussian blue. In fact, on panel A, the presence of the peak related to the carbon–nitrogen bond at ca. 2088 cm^{-1} in the ER-FTIR spectra suggested the use of Prussian blue (Figure 4) [27,28]. In this same area, the higher iron presence, along with the zinc signals in the XRF spectra, upholds a Prussian blue presence and confirms that it pertains to a modern restoration.

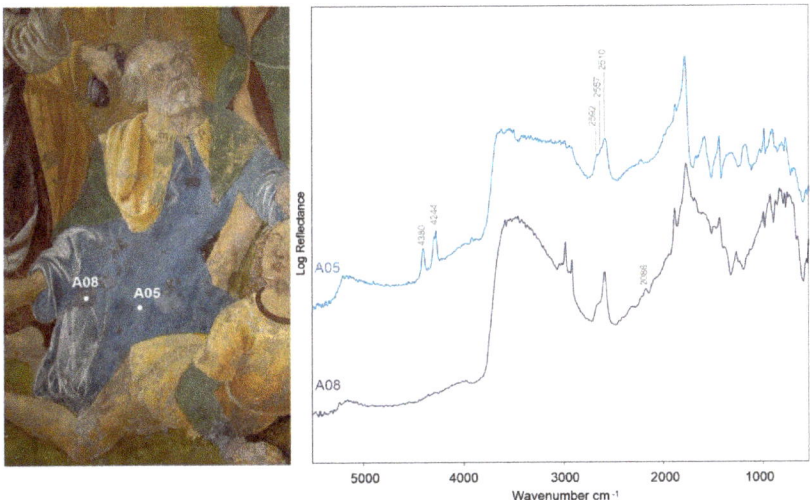

Figure 4. ER-FTIR spectra, acquired on a figure with the blue clothing on panel A, reveal the spectral evidence of azurite in the original area (A05) and Prussian blue in the restored part (A08).

Regarding the vault, instead, only XRF analyses were carried out, with the aim of investigating the composition of the blue background. However, due to problems relating to the height of the scaffold and the instrument placement, it was only possible to reach limited portions located in the lower part of the vault, while the top surfaces were not subjected to analysis. In all the spectra analyzed, as expected, high counts relating to calcium were observed. These are correlated to the Ca-rich matrix of the plaster. The calcium signals are always associated with those of strontium. Exceptions are the points analyzed in a single region of the vault (points V_08-V_12) in which the Sr signals are not detected; this could indicate that the calcium carbonate in that area was applied during restoration and shows a different geochemical signature. The XRF spectra also show intense counts relating to iron. The presence of this element, ubiquitously detected in the XRF spectra of all the areas of the vault analyzed, can be justified by the presence of a level of red preparation for the blue application, traditionally made with morellone [29]—an orange iron oxide and black carbon mixture applied on a calcite substratum. Copper signals are also always present, even if with very low intensities. These data allow us to hypothesize the presence of residual original azurite, which was subsequently integrated with modern materials. Regarding the identification of the blue pigment used in the repainting, a hypothesis can be formulated regarding the use of a mixture of Prussian blue and zinc white, the latter highlighted by the UVF images showing the lemon-yellow luminescence characteristic of the pigment [17], as already verified in the apsis panels. Indeed, as aforementioned, the presence of the Prussian blue pigment was confirmed by the ER-FTIR and XRF spectral results and by the analyses on cross-sections of the retouched blue backgrounds present in different areas of the pictorial cycle [30].

A further assumption that can be formulated for the points where there are no characteristic elements linked to a precise blue pigment (e.g., copper or cobalt) is the presence of a level of morellone, which could justify the intense signals of iron, with a superimposed layer made with ultramarine and lead white, applied to integrate the original azurite. Additionally, in this case, the hypothesis is partially confirmed by the results of the sampling and by the subsequent analytical investigations on the cross-sections, which highlight the presence of this mixture in some portions of the bottom of the vault. It is worth noting that, as previously stated, ultramarine was detected in some of the restored areas (individuated by the presence of high Zn signals in XRF spectra) of the apsis panels, for instance, in the blue vests in panel D.

3.3. Gildings

The gildings displayed a presence of gold X-ray characteristic lines, L_α and L_β, in the XRF spectra, respectively, at 9.71 keV and 11.44 keV, and also in those areas where they are no longer visible to the naked eye. Some of these areas, if observed by a digital microscope (see Figure 5), show the presence of residual islands of gildings.

Figure 5. Digital optical microscope images for residual gildings: from the left, panel A, panel D, and panel E. Red scale bars correspond to 0.2 mm.

On the gold gildings, the ER-FTIR revealed the use of lipidic material (i.e., oil), likely used to adhere the gold leaf to the surface. On the spectra acquired on all analyzed panels, sharp and intense signals with a derivative shape relative to $\nu(CH_2/CH_3)$ stretching appear at ca. 2930 and 2860 cm^{-1}, along with a weaker shoulder at ca. 2960 cm^{-1}, as well as the derivative band at ca. 1750 cm^{-1} due to the carbonyl asymmetric stretching band [14]. Moreover, in the near-infrared region, the peculiar doublet was detected due to the combination of methylenic C–H stretching and bending at ca. 4345 and 4255 cm^{-1} (Figure 3d) [31]. As highlighted by the multiband imaging investigations [17] in specific areas, the presence of organic material allowed us to map the original presence of the gildings, which were also on some dresses and veils. Indeed, all these areas are characterized by a yellow–orange fluorescence in the UVF (ultraviolet fluorescence photography) images [17] due to the presence of organic material and, namely, the lipidic fraction of the so-called missione used to make the gold adhere. The missione is the preparation (and sticking) layer of the gildings, made using linseed oil and small quantities of pigments. This type of mapping allowed us to recognize the former presence of gildings in several other areas, such as in the rays that leave the angels in the Resurrection scene (panel D).

3.4. Organic Materials

The ER-FTIR spectroscopy proved particularly effective in unveiling the types of organic materials occurring on the wall paintings examined. Spectral features suggested, in fact, the rather ubiquitous presence of lipid materials (oils and/or waxes) along with less widespread resins.

In addition to the gildings, traces of oils were detected in most of the spectra acquired on panels A, E, and D, highlighted by the bands at ca. 4345, 4255, 2960, 2930, 2860, and 1750 cm^{-1} (Figure 6).

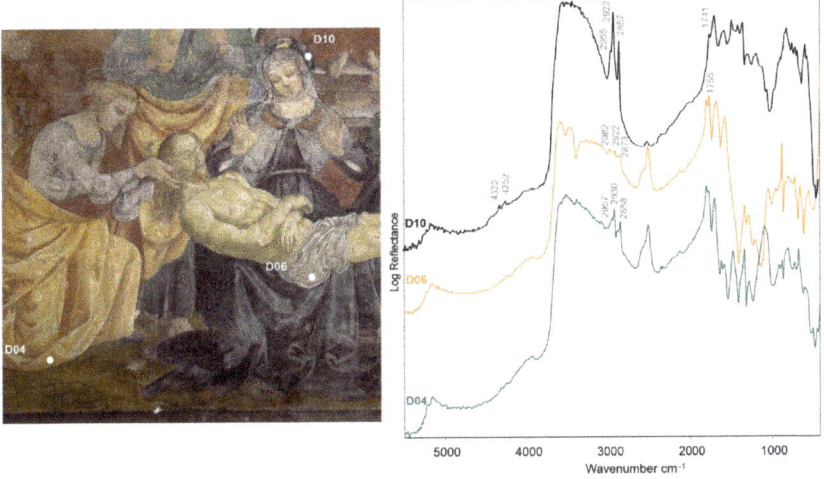

Figure 6. Detail of panel D, where oil (D10), resin (D06), and wax (D04) were detected.

Occasionally, instead, the spectral evidence of waxes and resins was observed. The waxes were easily distinguished by diagnostic CH$_2$ derivative-like stretching doublets at ca. at 2935 and 2858 cm^{-1}, with a weaker band near 2955 cm^{-1} (for the CH$_3$ stretching vibration, see Figure 6). The derivative effect at ca. 1740 cm^{-1} (C=O stretching vibration) was only occasionally observed as a very weak band. Where it lacks, the presence of paraffin could be hypothesized. Resins, instead, are featured by three broader, medium-weak signals at ca. 2980, 2950, and 2850 cm^{-1}, although the most diagnostic band is visible at ca. 1750–1740 cm^{-1}, related to the C=O stretching vibrations [22,23,32].

3.5. Alteration Products

The ER-FTIR also revealed the presence of specific alteration products affecting the wall paintings due to the alteration of the Ca-carbonate substrate and organic materials.

Most spectra from panels A, D, and E, in fact, showed the spectral evidence of calcium sulfate, featured by the broad bands at ca. 2230 ($2\nu_3$ SO$_4$; $\nu_2 + \nu_L$ H$_2$O) and 2140 cm^{-1} ($\nu_1 + \nu_3$ SO$_4$) and peaks at ca. 1150 cm^{-1} (ν_3 antisymmetric SO$_4$ stretching vibration modes), 670, and 600 cm^{-1} (ν_4 antisymmetric SO$_4$ bending vibration modes), along with the O-H stretching bands between 3600 and 3400 cm^{-1} (Figure 7a) [20].

Calcium sulfate could have several origins, as ion sulfates can be present in the form of impurities in the plaster, or calcium sulfate can arise from the soil; sulfate species are dissolved in soil water and, for capillary action, migrate and crystallize on wall surfaces [33]. An alternative hypothesis linked to the high presence of retouched areas is that calcium sulfate comes from the pigments used by restorers, as most of the time, calcium sulfate was added to the rated pigment. Indeed, sub-efflorescence and efflorescence, with consequent detachment and disintegration, were observed on the restoration mortars near the windows of panels A and C and along the lower part of the apsidal conch [34]. It is also worth noting that calcium sulfate was also detected in the samples from the Saturnino Gatti cycle through micro-stratigraphic analyses of thin sections [35] and the characterization of sampled soluble salts [36].

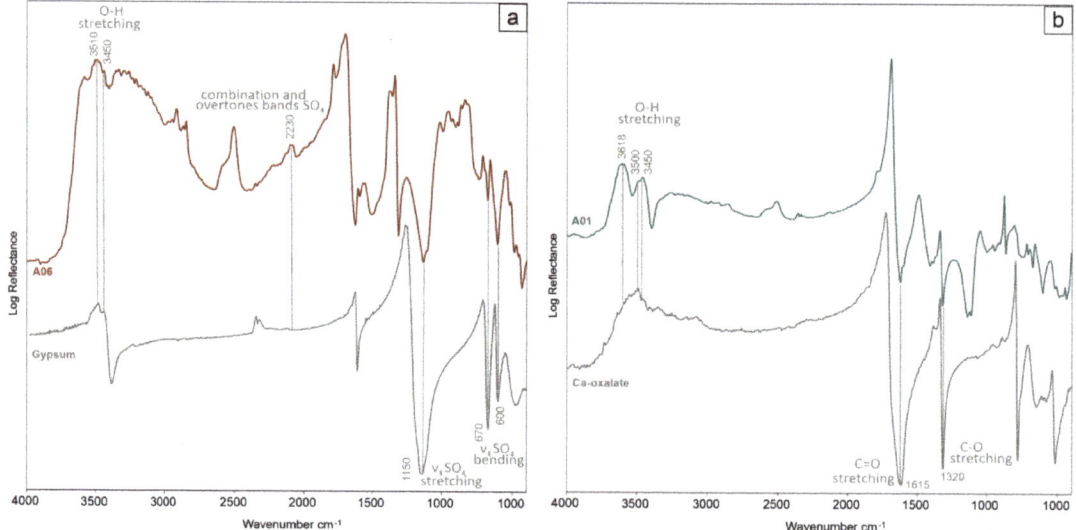

Figure 7. ER-FTIR spectra acquired on panel A show the spectral evidence of Ca-sulfates (**a**) and Ca-oxalates (**b**). Reference spectra [22] were also reported for comparison.

Moreover, in most spectra, the characteristic bands of calcium oxalate at ca. 1615 cm^{-1} (C=O vibration) and 1320 cm^{-1} (C-O vibration) were observed (Figure 7b) [22].

4. Conclusions

This paper presents one of the research activities in the frame of the study project of the pictorial cycle attributed to Saturnino Gatti at the church of San Panfilo in Villagrande di Tornimparte. The analytical campaign was promoted and coordinated by the AIAr (Italian Archaeometry Association) and established in cooperation with the Italian Ministry of Culture prior to the restoration works that became strictly necessary after the 2009 great earthquake in the region. In this context, we have presented the results obtained using in situ noninvasive analyses performed after, and guided by, the imaging techniques with the aim of characterizing the original and superimposed materials and, thus, allowing minimization of the sampling areas for further laboratory investigations.

The optimized analytical approach proposed, based on noninvasive diagnostics, proved to be powerful enough to characterize the present materials, ranging from the original pictorial layers to repainting and retouching, and to identify the techniques employed for the realizations of the wall paintings. The combination of elemental analysis with molecular characterization, provided by X-ray fluorescence and UV–vis spectroscopies, respectively, allows for the in situ noninvasive acquisition of a remarkable amount of information about the paint materials used by Saturnino Gatti. His palette comprises pigments, either used pure or in a mixture, to create different hues and are all compatible with the coeval pictorial technique. For the flesh tones, the artist used a mixture of calcium-based white with ochres/earths and amounts of vermillion. The presence of lead white has been, instead, highlighted in the retouched areas. Azurite was identified as the original pigment for the blue hues, while restorations occurred, probably at different times while using ultramarine or Prussian blue. Green earth, copper-based green, and chromium-based green were identified in different points or mixtures and probably pertained to the different interventions in the artwork. For the red hues, Saturnino Gatti painted mainly using iron oxides; in a few cases, these were mixed or in superimposition with vermilion. For the yellow and brown tones, the original areas revealed iron hydroxides.

The noninvasive approach suggests the use of a gilding technique with a golden leaf adhered to a red bolus preparation in areas where it is no longer visible to the naked eye.

Moreover, in addition to the original palette, the analytical protocol identified modern paint materials, such as Prussian blue, zinc white, copper, and chromium-based greens, used for numerous retouching. The repainted areas analyzed are probably attributable to several conservation interventions carried out in different periods.

In conclusion, a strong synergy with the imaging techniques survey performed on the same pictorial cycle, allowed an implementation on the quality of the punctual analyses results, demonstrating how a well-designed noninvasive campaign permits to drastically reduce a sampling, necessary to complete the information on the materials and the artist's technique.

Author Contributions: Investigation, L.B., S.C., A.G., C.G., F.I. and D.M.; Data Curation, L.B., S.C., A.G., C.G., F.I. and D.M.; Visualization: L.B., C.G. and D.M.; Writing—Original Draft Preparation, L.B., C.G. and D.M.; Writing—Review and Editing, L.B., S.C., A.G., C.G., F.I. and D.M.; Supervision, A.G. All authors have read and agreed to the published version of the manuscript.

Funding: This work was performed in the framework of the Project Tornimparte—"Archeometric investigation of the pictorial cycle of Saturnino Gatti in Tornimparte (AQ, Italy)", sponsored in 2021 by the Italian Association of Archeometry, AIAR (www.associazioneaiar.com (accessed on 27 May 2023)).

Data Availability Statement: The data presented in this study are available on request from the Italian Association of Archeometry, AIAR (www.associazioneaiar.com (accessed on 27 May 2023)).

Acknowledgments: The authors would like to thank the Proloco Tornimparte Association and the mayor of the municipality of Tornimparte for the availability and support shown during the diagnostic campaign.

Conflicts of Interest: The authors declare no conflict of interest.

References

1. Galli, A.; Alberghina, M.F.; Re, A.; Magrini, D.; Grifa, C.; Ponterio, R.C.; La Russa, M.F. Special Issue: Results of the II National Research project of AIAr: Archaeometric study of the frescoes by Saturnino Gatti and workshop at the church of San Panfilo in Tornimparte (AQ, Italy). *Appl. Sci.* **2023**. to be submitted.
2. Idjouadiene, L.; Mostefaoui, T.A.; Naitbouda, A.; Djermoune, H.; Mechehed, D.E.; Gargano, M.; Bonizzoni, L. First Applications of Non-Invasive Techniques on Algerian Heritage Manuscripts: The LMUHUB ULAHBIB Ancient Manuscript Collection from Kabylia Region (Afniq n Ccix Lmuhub). *J. Cult. Herit.* **2021**, *49*, 289–297. [CrossRef]
3. Bonizzoni, L.; Caglio, S.; Frezzato, F.; Martini, M.; Villa, V.; Galli, A. Balla's Bouquet: A Pigment Study for Flowers and Lights. *J. Cult. Herit.* **2021**, *52*, 164–170. [CrossRef]
4. Bonizzoni, L.; Bruni, S.; Galli, A.; Gargano, M.; Guglielmi, V.; Ludwig, N.; Lodi, L.; Martini, M. Non-Invasive in Situ Analytical Techniques Working in Synergy: The Application on Graduals Held in the Certosa Di Pavia. *Microchem. J.* **2016**, *126*, 172–180. [CrossRef]
5. Bonizzoni, L.; Bruni, S.; Gargano, M.; Guglielmi, V.; Zaffino, C.; Pezzotta, A.; Pilato, A.; Auricchio, T.; Delvaux, L.; Ludwig, N. Use of Integrated Non-Invasive Analyses for Pigment Characterization and Indirect Dating of Old Restorations on One Egyptian Coffin of the XXI Dynasty. *Microchem. J.* **2018**, *138*, 122–131. [CrossRef]
6. Bracci, S.; Cantisani, E.; Conti, C.; Magrini, D.; Vettori, S.; Tomassini, P.; Marano, M. Enriching the Knowledge of Ostia Antica Painted Fragments: A Multi-Methodological Approach. *Spectrochim. Acta Part A Mol. Biomol. Spectrosc.* **2022**, *265*, 120260. [CrossRef]
7. Bracci, S.; Cagnini, A.; Colombini, M.P.; Cuzman, O.A.; Fratini, F.; Galeotti, M.; Magrini, D.; Manganelli del Fà, R.; Porcinai, S.; Rescic, S.; et al. A Multi-Analytical Approach to Monitor Three Outdoor Contemporary Artworks at the Gori Collection (Fattoria Di Celle, Santomato, Pistoia, Italy). *Microchem. J.* **2016**, *124*, 878–888. [CrossRef]
8. Bartolozzi, G.; Bracci, S.; Cantisani, E.; Iannaccone, R.; Magrini, D.; Picollo, M. Non-Invasive Techniques Applied to the Alchemical Codex of the State Archive of Florence. *Spectrochim. Acta Part A Mol. Biomol. Spectrosc.* **2020**, *240*, 118562. [CrossRef]
9. Pinna, D.; Bracci, S.; Magrini, D.; Salvadori, B.; Andreotti, A.; Colombini, M.P. Deterioration and Discoloration of Historical Protective Treatments on Marble. *Environ. Sci. Pollut. Res.* **2022**, *29*, 20694–20710. [CrossRef]

10. Titubante, M.; Marconi, C.; Citiulo, L.; Mosca Conte, A.; Mazzuca, C.; Petrucci, F.; Pulci, O.; Tumiati, M.; Wang, S.; Micheli, L.; et al. Analysis and Diagnosis of the State of Conservation and Restoration of Paper-Based Artifacts: A Non-Invasive Approach. *J. Cult. Herit.* **2022**, *55*, 290–299. [CrossRef]
11. Macro, N.; Ioele, M.; Cattaneo, B.; De Cesare, G.; Di Lorenzo, F.; Storari, M.; Lazzari, M. Detection of Bronze Paint Degradation Products in a Contemporary Artwork by Combined Non-Invasive and Micro-Destructive Approach. *Microchem. J.* **2020**, *159*, 105482. [CrossRef]
12. Casanova, E.; Pelé-Meziani, C.; Guilminot, É.; Mevellec, J.-Y.; Riquier-Bouclet, C.; Vinçotte, A.; Lemoine, G. The Use of Vibrational Spectroscopy Techniques as a Tool for the Discrimination and Identification of the Natural and Synthetic Organic Compounds Used in Conservation. *Anal. Methods* **2016**, *8*, 8514–8527. [CrossRef]
13. Doménech-Carbó, M.T.; Doménech-Carbó, A.; Gimeno-Adelantado, J.V.; Bosch-Reig, F. Identification of Synthetic Resins Used in Works of Art by Fourier Transform Infrared Spectroscopy. *Appl. Spectrosc.* **2001**, *55*, 1590–1602. [CrossRef]
14. Rosi, F.; Daveri, A.; Moretti, P.; Brunetti, B.G.; Miliani, C. Interpretation of Mid and Near-Infrared Reflection Properties of Synthetic Polymer Paints for the Non-Invasive Assessment of Binding Media in Twentieth-Century Pictorial Artworks. *Microchem. J.* **2016**, *124*, 898–908. [CrossRef]
15. Ludwig, N.; Orsilli, J.; Bonizzoni, L.; Gargano, M. UV-IR Image Enhancement for Mapping Restorations Applied on an Egyptian Coffin of the XXI Dynasty. *Archaeol. Anthropol. Sci.* **2019**, *11*, 6841–6850. [CrossRef]
16. Mannetti, T.R.; Chelli, N.; Vecchioli, G. *Saturnino Gatti nella Chiesa di San Panfilo a Tornimparte*; Ferri, A., Ed.; Edizioni del Gallo Cedrone: L'Aquila, Italy, 1992.
17. Bonizzoni, L.; Caglio, S.; Galli, A.; Lanteri, L.; Pelosi, C. Materials and technique: The first look at Saturnino Gatti. *Appl. Sci.* 2023, submitted.
18. Mercurio, M.; Rossi, M.; Izzo, F.; Cappelletti, P.; Germinario, C.; Grifa, C.; Petrelli, M.; Vergara, A.; Langella, A. The Characterization of Natural Gemstones Using Non-Invasive FT-IR Spectroscopy: New Data on Tourmalines. *Talanta* **2018**, *178*, 147–159. [CrossRef]
19. Bonizzoni, L.; Caglio, S.; Galli, A.; Poldi, G. Comparison of three portable EDXRF spectrometers for pigment characterization. *X-Ray Spectrom.* **2010**, *39*, 233–242. [CrossRef]
20. Germinario, C.; Francesco, I.; Mercurio, M.; Langella, A.; Sali, D.; Kakoulli, I.; De Bonis, A.; Grifa, C. Multi-Analytical and Non-Invasive Characterization of the Polychromy of Wall Paintings at the Domus of Octavius Quartio in Pompeii. *Eur. Phys. J. Plus* **2018**, *133*, 359. [CrossRef]
21. Briani, F.; Caridi, F.; Ferella, F.; Gueli, A.M.; Marchegiani, F.; Nisi, S.; Paladini, G.; Pecchioni, E.; Politi, G.; Santo, A.P.; et al. Multi-technique characterization of painting drawings of the pictorial cycle at the San Panfilo Church in Tornimparte (AQ). *Appl. Sci.* **2023**, *13*, 6492. [CrossRef]
22. Izzo, F.; Germinario, C.; Grifa, C.; Langella, A.; Mercurio, M. External Reflectance FTIR Dataset (4000–400 cm^{-1}) for the Identification of Relevant Mineralogical Phases Forming Cultural Heritage Materials. *Infrared Phys. Technol.* **2020**, *106*, 103266. [CrossRef]
23. Invernizzi, C.; Rovetta, T.; Licchelli, M.; Malagodi, M. Mid and Near-Infrared Reflection Spectral Database of Natural Organic Materials in the Cultural Heritage Field. *Int. J. Anal. Chem.* **2018**, *2018*, 1–16. [CrossRef]
24. Mazzinghi, A. XRF analyses for the study of painting technique and degradation on frescoes by Beato Angelico: First results. *Il Nuovo Cim.* **2014**, *37*, 253–262. [CrossRef]
25. Gutman Levstik, M.; Mladenovič, A.; Kriznar, A.; Dolenec, S. A Raman microspectroscopy-based comparison of pigments applied in two gothic wall paintings in Slovenia. *Period. Mineral.* **2019**, *88*, 95–104. [CrossRef]
26. Miliani, C.; Rosi, F.; Daveri, A.; Brunetti, B.G. Reflection Infrared Spectroscopy for the Non-Invasive in Situ Study of Artists' Pigments. *Appl. Phys. A* **2012**, *106*, 295–307. [CrossRef]
27. Kahrim, K.; Daveri, A.; Rocchi, P.; de Cesare, G.; Cartechini, L.; Miliani, C.; Brunetti, B.G.; Sgamellotti, A. The Application of in Situ Mid-FTIR Fibre-Optic Reflectance Spectroscopy and GC–MS Analysis to Monitor and Evaluate Painting Cleaning. *Spectrochim. Acta Part A Mol. Biomol. Spectrosc.* **2009**, *74*, 1182–1188. [CrossRef] [PubMed]
28. Rampazzi, L.; Brunello, V.; Corti, C.; Lissoni, E. Non-Invasive Techniques for Revealing the Palette of the Romantic Painter Francesco Hayez. *Spectrochim. Acta Part A Mol. Biomol. Spectrosc.* **2017**, *176*, 142–154. [CrossRef] [PubMed]
29. Scremin, B.F. Raman Study of a Work of Art Fragment. *Vib. Spectrosc.* **2018**, *99*, 162–166. [CrossRef]
30. Armetta, F.; Giuffrida, D.; Ponterio, R.C.; Falcon Martinez, M.F.; Briani, F.; Pecchioni, E.; Alba Patrizia Santo, A.P.; Ciaramitaro, V.C.; Saladino, M.L. Looking for the original materials and evidence of restoration at the Vault of the San Panfilo Church in Tornimparte (AQ). *Appl. Sci.* **2023**. to be submitted.
31. Vagnini, M.; Miliani, C.; Cartechini, L.; Rocchi, P.; Brunetti, B.G.; Sgamellotti, A. FT-NIR Spectroscopy for Non-Invasive Identification of Natural Polymers and Resins in Easel Paintings. *Anal. Bioanal. Chem.* **2009**, *395*, 2107–2118. [CrossRef]
32. Alberghina, M.F.; Germinario, C.; Bartolozzi, G.; Bracci, S.; Grifa, C.; Izzo, F.; Russa, M.F.L.; Magrini, D.; Massa, E.; Mercurio, M.; et al. The Tomb of the Diver and the Frescoed Tombs in Paestum (Southern Italy): New Insights from a Comparative Archaeometric Study. *PLoS ONE* **2020**, *15*, e0232375. [CrossRef] [PubMed]
33. Cultrone, G.; Arizzi, A.; Sebastián, E.; Rodriguez-Navarro, C. Sulfation of Calcitic and Dolomitic Lime Mortars in the Presence of Diesel Particulate Matter. *Environ. Geol.* **2008**, *56*, 741–752. [CrossRef]

34. Lanteri, L.; Calandra, S.; Briani, F.; Germinario, C.; Izzo, F.; Pagano, S.; Pelosi, C.; Santo, A.P. 3D Photogrammetric Survey, Raking Light Photography and Mapping of Degradation Phenomena of the Early Renaissance Wall Paintings by Saturnino Gatti—Case Study of the St. Panfilo Church in Tornimparte (L'Aquila, Italy). *Appl. Sci.* **2023**, *13*, 5689. [CrossRef]
35. Germinario, L.; Giannossa, L.C.; Lezzerini, M.; Mangone, A.; Mazzoli, C.; Pagnotta, S.; Spampinato, M.; Zoleo, A.; Eramo, G. Petrographic and chemical characterization of the frescoes by Saturnino Gatti (central Italy, 15th century): Microstratigraphic analyses on thin sections. *Appl. Sci.* **2023**. *to be submitted*.
36. Comite, V.; Bergomi, A.; Lombardi, C.A.; Fermo, P. Characterization of soluble salts on the frescoes by Saturnino Gatti in the church of San Panfilo in Villagrande di Tornim-parte (L'Aquila). *Appl. Sci.* **2023**. *to be submitted*.

Disclaimer/Publisher's Note: The statements, opinions and data contained in all publications are solely those of the individual author(s) and contributor(s) and not of MDPI and/or the editor(s). MDPI and/or the editor(s) disclaim responsibility for any injury to people or property resulting from any ideas, methods, instructions or products referred to in the content.

Article

Looking for the Original Materials and Evidence of Restoration at the Vault of the San Panfilo Church in Tornimparte (AQ)

Francesco Armetta [1,2], Dario Giuffrida [2], Rosina C. Ponterio [2], Maria Fernanda Falcon Martinez [3], Francesca Briani [4], Elena Pecchioni [5], Alba Patrizia Santo [5], Veronica C. Ciaramitaro [1] and Maria Luisa Saladino [1,2,*]

[1] Dipartimento Scienze e Tecnologie Biologiche, Chimiche e Farmaceutiche, Università di Palermo, Viale delle Scienze Bld. 17, 90128 Palermo, Italy; francesco.armetta01@unipa.it (F.A.); veronicaconcetta.ciaramitaro@unipa.it (V.C.C.)
[2] CNR—Istituto per i Processi Chimico Fisici, Viale Ferdinando Stagno D'Alcontres 37, 98158 Messina, Italy; dario.giuffrida@ipcf.cnr.it (D.G.); ponterio@ipcf.cnr.it (R.C.P.)
[3] Soprintendenza Belle Arti, Archeologia e Paesaggio per le Province di L'Aquila e Teramo, Via San Basilio, 67100 L'Aquila, Italy; mariafernanda.falconmartinez@cultura.gov.it
[4] Adarte S.n.c., Via Agnoletti Enriques Annamaria 3, 50141 Firenze, Italy; francescabriani@adartesnc.it
[5] Dipartimento di Scienze della Terra, Università degli Studi di Firenze, Via La Pira 4, 50121 Firenze, Italy; elena.pecchioni@unifi.it (E.P.); alba.santo@unifi.it (A.P.S.)
* Correspondence: marialuisa.saladino@unipa.it

Abstract: This paper reports the investigation of six microsamples collected from the vault of the San Panfilo Church in Tornimparte (AQ). The aim was to detect the composition of the pigments and protective/varnishes, and to investigate the executive technique, the conservation state, and the evidence of the restoration works carried out in the past. Six microsamples were analyzed by optical microscopy, scanning electron microscopy coupled with energy-dispersive spectroscopy (EDS), X-ray fluorescence (XRF), and infrared and Raman spectroscopy. The investigations were carried out within the framework of the Tornimparte project "Archeometric investigation of the pictorial cycle of Saturnino Gatti in Tornimparte (AQ, Italy)" sponsored in 2021 by the Italian Association of Archeometry (AIAr).

Keywords: pigments; retouching; executive technique; conservation state; XRF; FT-IR; SEM

Citation: Armetta, F.; Giuffrida, D.; Ponterio, R.C.; Martinez, M.F.F.; Briani, F.; Pecchioni, E.; Santo, A.P.; Ciaramitaro, V.C.; Saladino, M.L. Looking for the Original Materials and Evidence of Restoration at the Vault of the San Panfilo Church in Tornimparte (AQ). *Appl. Sci.* **2023**, *13*, 7088. https://doi.org/10.3390/app13127088

Academic Editor: Vittoria Guglielmi

Received: 15 April 2023
Revised: 10 June 2023
Accepted: 10 June 2023
Published: 13 June 2023

Copyright: © 2023 by the authors. Licensee MDPI, Basel, Switzerland. This article is an open access article distributed under the terms and conditions of the Creative Commons Attribution (CC BY) license (https://creativecommons.org/licenses/by/4.0/).

1. Introduction

This paper contributes to the Special Issue "Results of the II National Research project of AIAr: Archaeometric study of the frescoes by Saturnino Gatti and workshop at the church of San Panfilo in Villagrande di Tornimparte (AQ, Italy)", in which the scientific results of II National Research Project conducted by members of the Italian Association of Archaeometry (AIAr (Associazione Italiana di Archeometria. http://www.associazioneaiar.com/wp/home-2/, accessed on 20 February 2023)) are discussed and collected. For in-depth details on the aims of the project, see the introduction of the Special Issue [1]. This work is a first attempt to understand the historical past of these frescoes on the basis of a direct observation of the current state of conservation and a study of the (fragmentary) documentation preserved in the archive [2,3]. The frame in which the execution of the pictorial cycle can be placed begins with a note of an advanced payment given to the artist in 1489; it was 1491 when Saturnino started to work on the frescoes which were completed in 1494 [2]. Art historians consider this cycle of frescoes the cornerstone of Saturnino Gatti's activity as a painter; it is also considered his masterpiece for the close and evident similarities that the frescoes demonstrate with the models of his famous master, the Florentine sculptor and painter Andrea del Verrocchio. New information from the recent analysis campaign is added [4–6]. The pre-restoration analysis is an invaluable contribution, setting objective data on the quality and nature of the original and restoration materials and degradation processes, leading us on a parallel path that confirms and enriches the information reported on the documents in our possession. In the long life of an artwork, the conservation events are difficult to reconstruct [7,8], on one hand, because the news is often fragmentary and inaccurate, and, on the other, because it is not possible to see beyond the mere aesthetic aspect through only visual inspection. In addition to the normal degradation to which the constituent materials are subjected, we have to deal with clumsy and invasive interventions and inadequate maintenance work. In this regard, there is not much to add to the description made by Giuseppina Vecchioli where she describes how the frescoes of the vault appear "grossly repainted", and how the figures are excessively reintegrated [9]. This is how the most important pictorial cycle of Saturnino Gatti appears today. Its history is not only made up of localized and limited interventions responding to the scarce economic resources of the moment, since other factors have determined uneven intervention, always characterized by different criteria. It is then necessary, as spectators, to take a step back and try to take a view of the most important events that have marked the history of these frescoes. Through the examination of some documents, we can verify how there have been episodes causing degradation in paintings over the years. Some of these events are listed in a letter signed by Mons. Bernardino Santucci dated 12 February 1964, and addressed to the Superintendent of Medieval and Modern Art and the Minister of Education: "the earthquake of 15 January 1915, devastating war of 1940–1944, very hard winter of 1962–1963, and permanent moisture penetrating from the roofs and the floor under which are burials of the dead" (12 February 1964). The first documented information placed in the records of this Superintendency concerning this pictorial cycle dates back to 1950 and after the note signed by the Superintendent which describes: "Some bombs that exploded nearby damaged the roof and the windows in various points", attaching an appraisal of the amount of 40,000 ITL "to prevent the next winter season from damaging the paintings". Another note of 3 August 1950 for the financing "will follow ... on the chap. 257 relating to war damages". The total amount of 3,400,000 ITL concerned three of the most damaged sacred buildings in the area, including S. Panfilo di Tornimparte (Letter from Soprintendenza ai Monumenti e Gallerie degli Abruzzi e Molise, addressed to Ministero della Pubblica Istruzione e Direzione Generale delle Belle Arti, 3 August 1950). To the catastrophic events are added others dictated by the lack of maintenance and neglect (such as the fire of October 1958), or little availability of funds, all leading us to the same result, i.e., "the frescoes of Gatti are deteriorating visibly". A letter written by the then parish priest Bernardino Santucci recalls the damage suffered by the building: "After the restorations of 1929, nothing more remarkable has been done for the conservation and decoration of the Holy Temple and its sacred works of art of which it is rich". It continues: "Because of the bad weather of the mountain and of the war, everything is in a state of decay and ruin and urgent action is required as the pictorial cycle continues to deteriorate" (Letter from Sac. Bernardino Santucci, addressed to Direttore Generale, 15 February 1951). At the previous request, and, after some exchanges between the superintendency and the Ministry of Public Education, in August, the Superintendent requested and communicated that it would be sufficient to carry out the work on the frescoes, that another estimate was received, that an allocation was approved, and that the announcement of the forthcoming start of the work was given (Letter from Soprintendente Dott. Arch. Umberto Chierici, addressed to Reverendo Parroco di S. Panfilo, 13 September 1951). In 1951, the allocation of the necessary funds for the restoration of the frescoes by Saturnino Gatti in Tornimparte was approved. For the approval of these funds, an expense report is found in the documents that describe in detail the intervention proposed on 30 August 1951: "cleaning of the painted surface and removal of vast in paintings altered by time, fixing of the crumbling and crumbling color and consolidation of swollen plaster areas with injections of lime caseate, filling of cracks or gaps in backgrounds with local glazes".

The Wall Paintings of the Apse of San Panfilo

These paintings were restored for the first time in 1929 by Domenico Brizi. Another intervention to mention is that by Enrico Vivio, following the fire of 1958–1959. A new communication dated 5 November 1958 resumes the need for a restoration of the pictorial cycle, this time due to the fire caused by electric damage in the night between 5 and 6 October in the parish sacristy, with consequent damage to the frescoes in the apse [5]. On 12 November 1958, the Superintendent announced that a technician had been sent to carry out the necessary investigations, confirming in a subsequent note dated 12 November 1956 the start of "cleaning and restoration" work and confirmed by a letter of thanks dated 14 April 1959, written by the parish priest of San Panfilo Mons. Bernardino Santucci (Letter from Mons. Bernardino Santucci, addressed to Soprintendente all'arte Medievale e Moderna L'Aquila, 10 April 1959). About 15 years after this last restoration, a new overall intervention of the paintings was entrusted to the restorers Antonio Liberti and Umberto Marini, and, in a letter addressed to B. Santucci, the architect Mario Moretti specifies that the intervention would begin, but only once the roof was restored " . . . and, nevertheless, not before the spring of 1971" (29 December 1970). "The frescoes of the vault, due to the infiltration of moisture from the roof, have large oxidations and blooms of carbonates and nitrates and the presence of mildew, which have seriously altered the blues. In some areas the color has fallen; in others, the color is curled and unsafe" (arch. Mario Moretti, 17 April 1972). The architect Moretti describes a very bad state of conservation of the paintings of the vault. Among other things, he mentions the serious alteration "of the blues", covering a large area; hence, at the time, a trace of the original coloring of the sky was perhaps still visible. It is worth mentioning that, with the cleaning carried out on that occasion, according to Liberati, the original colors were recovered as much as possible, freeing the frescoes of the superfetations. According to the judgment of Ferdinando Bologna, the cleaning was excessive, compromising the legibility of important details. Unfortunately, details of the appraisal presented are sparse and do not offer useful information to trace the cleaning system used or if a pre-consolidation work of the pictorial film was performed before cleaning the surface, deducing that the frequent infiltrations had generated a serious problem of decohesion of the pictorial layer. At first glance, the pictorial work still manages to convey the beauty of the composition and the accurate stroke of its figures. Conducting a more accurate visual inspection, the numerous "interferences" against the pictorial film creep into the eyes of the viewer; the celestial vault appears covered by a heavy and compact dark-blue application and many interventions on some figures, distorted by interpretative integrations. On the walls, there are gaps, abrasions of the pictorial film and humidity problems. The investigations conducted, using different methods, analyzed each situation and found answers that are not handed down in the documents. In the apse, the one that first catches the eye is the background of a very compact blue. The sky is heavily repainted, leaving a set of figures "cut out and outlined" by heavy coloring. On the other hand, the figures floating in the sky show both signs of degradation, such as abrasions and small gaps, and repaints that overload the features of hands and faces. The succession of winter seasons, frost and thaw, and repeated episodes of infiltration from the roof have caused the degradation of the supporting plaster and serious decohesion of the pictorial film as described by the architect Mario Moretti in his report. However, it is to be assumed that these episodes have been a constant in the long life of these paintings, as demonstrated by the same repainting already mentioned. Therefore, the aim of this work is to identify the pigments, binders, and protective/varnishes, to study the stratigraphy, and to investigate the executive technique and the conservation state of the paintings at the Vault. Attention was paid to evaluate the original pigments or eventually the ones used during a restoration work performed in the past. Six microsamples collected in the Vault were analyzed with the main aim of understanding the stratigraphy and providing a *post quem* term for the realization of the extensive integrations and additions, which altered the colors and original reading of the fresco. The adopted methodology is based on a combined use of several techniques, widely used for elemental and molecular investigations of pigments in the field of Cultural Heritage: observation of cross-sections under optical microscopy and scanning electron microscopy coupled with EDS detection, X-ray fluorescence (XRF), and Raman spectroscopy, along with an investigation of each layer thought µ-infrared spectroscopy [10–12].

2. Materials and Methods

2.1. Materials

The six microsamples were selected on the basis of the results given by the multispectral investigation, reported in [4–6]. All microsamples were collected in the upper part of the vault, as reported in Figure 1.

Figure 1. Area of the vault where the microsamples have been collected. The number corresponds to the ID microsamples.

The ID samples and a photo of each microsample, together with the aim of the specific investigation, are reported in Table 1.

2.2. Methods

The cross-sections were prepared by embedding the collected microsamples into epoxy resin (bicomponent Prochima E30) and curing until solidification at room temperature. The resin was then cut and polished to get the stratigraphy across the center of the objects.

Optical microscopy observation was performed using a LEICA EZ24W optical microscope in reflection mode. Sample 39B cross-section was observed in reflected light by means of a ZEISS Axio Scope A1 microscope, equipped with a video camera (5 MP resolution), and image analysis software AxioVision (V1).

Table 1. ID sample, area of the sampling, photo of the microsamples, and their macroscopic description.

ID Sample	Photo of the Sampling Area (Indicated by the Yellow Arrows)	Photos of the Microsample (Front)
SG_37B		Gray pigment on degraded pictorial layer
SG_38		Green pigment on preparation layer
SG_39A		Blue pigment on red pictorial layer and then on plaster
SG_39B		Blue pigment on red pictorial layer and then on plaster
SG_41A		Violet pigment on plaster
SG_42		Dark-blue pigment

SEM investigation coupled with energy-dispersive X-ray spectroscopy (EDS) analysis was performed using a Phenom Pro X, Phenom-World (The Netherlands) with an optical magnification range of 20×–135×, electron magnification range of 80,000×–130,000×, maximal digital zoom of 12×, and acceleration voltage of 15 kV, as well as an EDS detector, with a nominal resolution of 10 nm or less. The microscope was equipped with a temperature-controlled (25 °C) sample holder. The samples were positioned on an aluminum stub using adhesive carbon tape. Morphological and semiquantitative microchemical analyses of sample 39B were performed using a SEM-EDS electronic microscope (ZEISS EVO MA 15) with a W-filament equipped with an analysis system in energy dispersion EDS/SDD, Oxford Ultimax 40 (40 mm^2 with resolution 127 eV @ 5.9 keV) and the Aztec 5.0 SP1 software. Measurements were performed on a carbon-metallized cross-section of the sample under the following operating conditions: accelerating potential of 15 kV, 500 pA beam current, working distance between 9 and 8.5 mm, 20 s live time as an acquisition rate useful for archiving at least 600,000 cts, on co-standard and process time 4 for point analyses, and 500 µs pixel dwell time for recording 1024 × 768 pixel resolution maps. The software for the microanalysis was Aztec 5.0 SP1 software, which uses the XPP matrix correction scheme developed by Pouchou and Pichoir in 1991 [13].

XRF spectra were acquired using a Tracer III SD Bruker AXS portable spectrometer. The irradiation by a Rhodium Target X-Ray tube operating at 40 kV and 11 µA and the detection of fluorescence X-rays by a 10 mm^2 silicon drift X-Flash detector allowed the detection of elements with atomic number Z > 11. A window of 3–4 mm in diameter determined the sampled area. Each spectrum was acquired for 30 s. The S1PXRF® software (https://s1pxrf.software.informer.com/, accessed on 14 April 2023) was used for data acquisition and spectral assignments. The fluorescence signal area was estimated once the deconvolution of the whole spectrum was performed using the software ARTAX 7. Ar, Ni, Pd, and Rh signals, due to the atmosphere and instrumental components, were also present in all spectra.

Raman spectra were acquired through a LabRAM HR-800 Jobin-Yvon spectrometer equipped with a laser at 633 nm, and a Bravo Bruker spectrometer equipped with two lasers at 785 and 853 nm.

The µ-IR spectra were acquired through a micro FT-IR Lumos Bruker spectrophotometer equipped with a Platinum ATR unit using a germanium crystal operating in the spectral range between 4000 and 600 cm^{-1}, with a spectral resolution of 2 cm^{-1} and 120 scans. The size of the investigated area was determined by the size of the crystal tip (~50 microns). In all spectra, a baseline correction of scattering was made. Data analysis was performed using the OPUS 7.5® software.

3. Results and Discussion

The ID samples, the macro-photos of the cross sections observed under the optical microscope, and the description of the observed layers are reported in Table 2.

For all samples, the plaster made of calcium carbonate ($CaCO_3$), consisting of aggregates of variable grain size and different colors, and the presence of a protective copolymer of vinyl-acrylate acetate/Mowilith DM912 [14], ascribed to a previous restoration, were identified. In most of the microsamples, the presence of weddelite or whewellite, attributable to the growth of microorganisms responsible for biological degradation [15], was identified.

The obtained results for each microsample are reported separately below.

3.1. SG_37B Microsample

This microsample is constituted by two layers:

- Plaster made with a light-colored binder (calcium carbonate, $CaCO_3$), consisting of aggregates of variable grain size and color;
- Continuous and homogeneous dark-gray layer.

Table 2. ID sample, macro-photos of the cross-section observed under the optical microscope, and description of the observed layers.

ID Sample	Macro-Photos of the Cross-Section Observed under the Optical Microscope	Description of the Observed Layers -from the Top to the Bottom-
SG_37B		2 layers: - Dark gray (mixture of organic black and ochre) - Plaster
SG_38		3 layers: - Green pigment - Grey primer (mixture of black, ochre, and white pigments) - Plaster
SG_39A		6 layers: - Blue pigment, retouching - Red pigment, retouching - White primer - Red pigment - White primer - Plaster
SG_39B		3 layers: - Blue pigment with white grains - Red pigment - Plaster
SG_41A		3 layers: - Red-violet pigment and gypsum - White primer - Plaster
SG_42		4 layers: - Green pigment - Blue pigment applied a secco - Red-violet pigment and gypsum - White primer - Plaster

The XRF spectra show the peaks of calcium (Ca), iron (Fe), and lead (Pb), suggesting that the gray layer is made up of a mixture of several parts of ochres (iron oxides) and a black pigment (such as carbon black), probably the so-called *caput mortuum* pigment, also known as *morellone* (a mixture of red ochre and organic black). The presence of lead (Pb) indicates the use of white lead pigment (Figure 2). The FT-IR spectrum shows absorption bands at 1783, 1428, 870, and 716 cm^{-1}, attributed to the characteristic vibrations of the carbonate ion (CO_3^{2-}) [16], used as a binder, and the absorption bands at 2973, 2926, 1731, 1433, 1370, 1227, 1120, 1020, 945, 865, 795, 632, and 604 cm^{-1} ascribable to a synthetic resin known as Mowilith DM 912 [17].

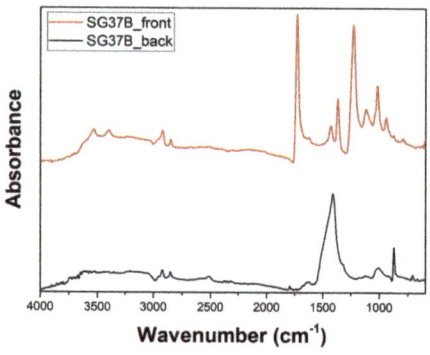
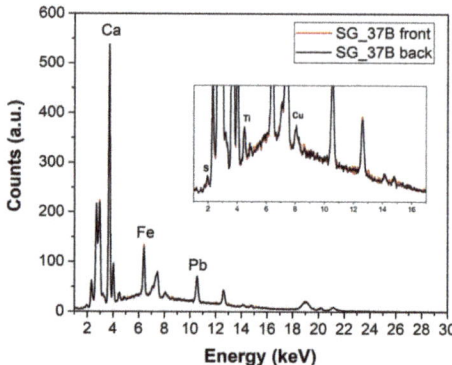

Figure 2. (**left**) FT-IR and (**right**) XRF spectra acquired on the front and back of the SG_37B micro sample before the preparation of the cross-section.

3.2. SG_38 Microsample

This microsample is constituted by three layers:

- Plaster made with a light-colored binder (calcium carbonate), consisting of aggregates of variable grain size and color;
- Continuous gray layer;
- Green layer.

The XRF spectra (Figure 3) show the peaks of calcium (Ca) and iron (Fe), suggesting that the gray layer was obtained by mixing a white pigment with a black pigment probably obtained from natural earth (organic black, a mixture of clay and calcium carbonate, iron, and manganese). For the green pigment, the presence of iron (Fe) and silicon (Si) in the XRF spectra suggests the use of green earth (ferrous and ferric silicates of potassium, manganese, and aluminum plus oxides of Fe, Mg, Al, and K). As with the previous microsample, the FT-IR analysis showed the presence of calcium carbonate and a synthetic resin known as Mowilith DM 912. The bands of whewellite (1623, 1320, and 780 cm^{-1}) were also present [15].

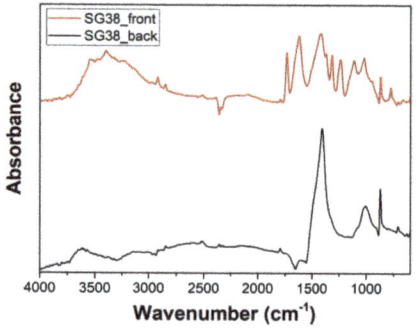
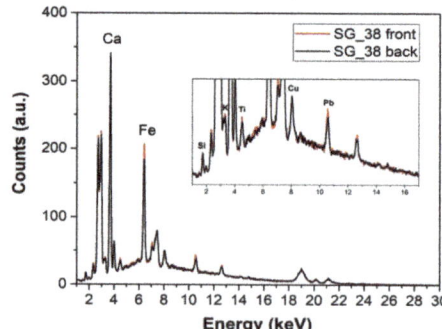

Figure 3. (**left**) FT-IR and (**right**) XRF spectra acquired on the front and back of the SG_38 microsample before the preparation of the cross-section.

3.3. SG_39A Microsample

This microsample is constituted by six layers:

- Plaster made with a light-colored binder (calcium carbonate);
- First layer of white drafting;
- Second layer of red drafting (red ochre/hematite);
- Third layer of white drafting (calcite);
- Red layer applied a secco, retouching;
- Blue layer (Prussian blue) applied a secco, retouching.

The EDS analysis, acquired along the dashed line reported in the SEM micrograph (Figure 4), showed the elements present in the different layers. The results mainly indicated the presence of calcium (Ca), lead (Pb), and iron (Fe). From the distribution of the elements along the cross-section of the sample, from the plaster toward the surface pictorial layers, a high Ca content can be observed, which is attributable to the calcite present in the plaster and in the first white layer, followed by Fe in correspondence of the second red layer, which suggests the use of red ochre. The use of minium or other lead-based pigments such as litharge or massicot cannot be excluded [18]. However, the presence of lead (Pb) may also indicate the use of lead white pigment. An increase in Ca was observed again, in correspondence with the third white layer, followed by the Fe presence for the second red layer and the last blue layer, which suggests the use of red ochre and Prussian blue ($Fe_4[Fe(CN)_6]_3$), respectively.

As for the previous microsamples, the FT-IR analysis showed the presence of calcium carbonate, a synthetic resin known as Mowilith DM 912, and whewellite and weddellite. In addition, the IR bands at about 2100 cm^{-1}, attributable to the stretching of the triple bond –C≡N. and two Raman bands at 520 and 2148 cm^{-1}, attributable to the ferric hexacyanoferrate, $Fe_4[Fe(CN)_6]_3$, are ascribable to the Prussian blue pigment. The Raman band at 1008 cm^{-1} is typical of gypsum ($CaSO_4$)·$2H_2O$. The presence of Prussian blue is ascribable to a retouching, applied a secco (Figure 5).

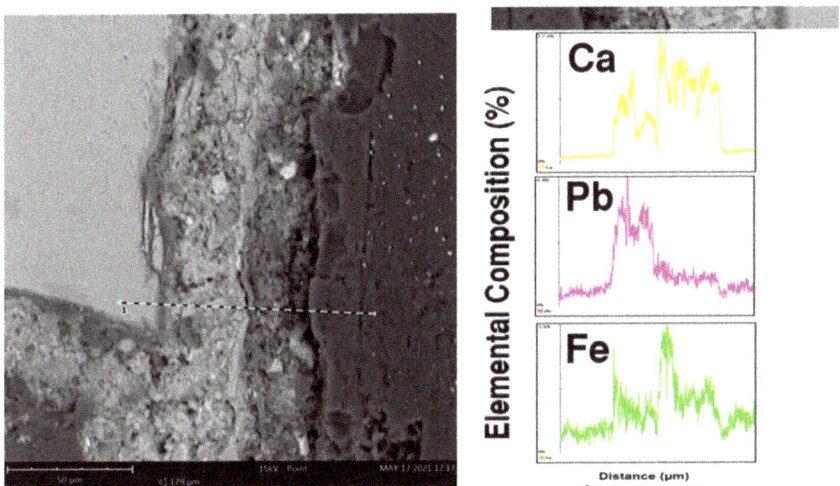

Figure 4. (**left**) SEM micrograph and (**right**) the corresponding EDS profile of SG_39A cross-section. The EDS investigation was performed along the dashed line to evaluate the distribution of Ca, Pb, and Fe from the back to front of the sample.

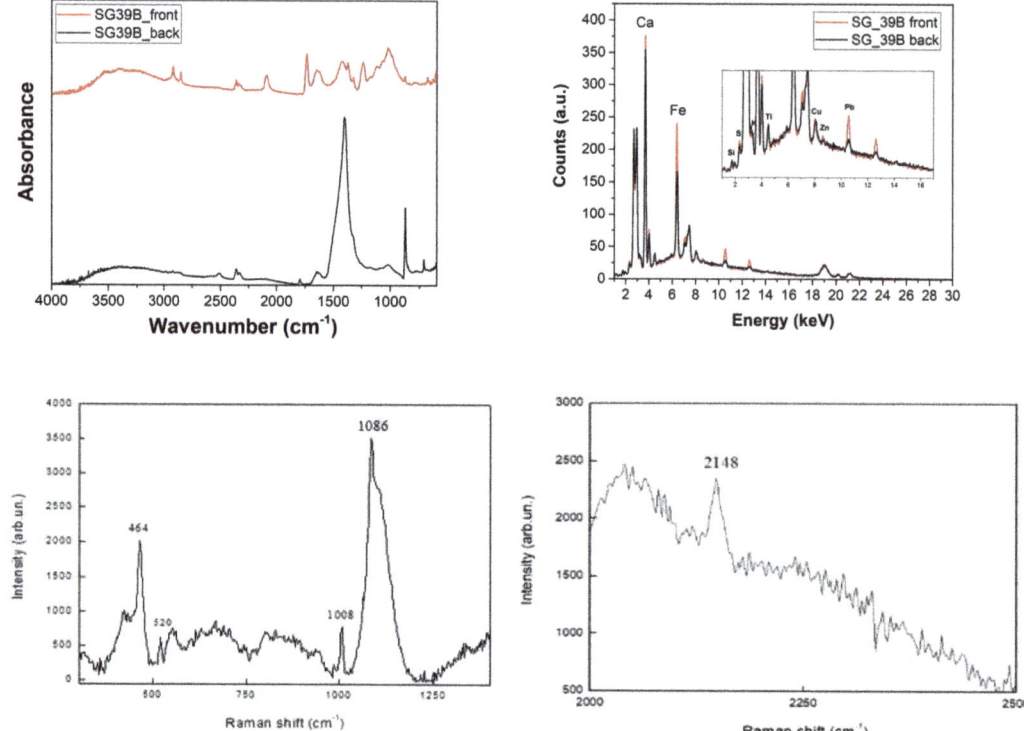

Figure 5. (**top left**) FT-IR and (**top right**) XRF spectra acquired on the front and back of the SG_39A microsample. (**bottom left, right**) Raman spectra acquired on SG_39A before the preparation of the cross-section.

3.4. SG_39B Microsample

This microsample is constituted by three layers:
- Plaster containing rounded and squared grains;
- First layer of red coating, probably the so-called *morellone or caput murtuum* (a mixture of red ochre and organic black) [19];
- Blue layer (approximately 20 µm) with white grains inside.

The EDS analysis was acquired on the three layers and, when required, in correspondence with specific colored grains, in order to characterize every single pigment (Figure 6). The acquired spectrum on the observed underlayer (plaster) showed large amounts of calcium (Ca), in addition to silicon (Si), aluminum (Al), sodium (Na), magnesium (Mg), potassium (K), and iron (Fe). Concerning the above red layer (SEM-EDS not reported), the only detected chromophore element was Fe, in relation to the use of red ochre, which supports the hypothesis made for *morellone*. The analysis of the blue layer showed the presence of a high percentage of lead (Pb), in white grains, due to the use of lead white pigment, as well as Si, Al, Na, and sulfur (S), in blue grains, all distinctive elements for ultramarine blue.

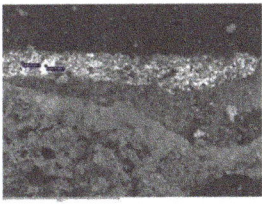 Sample 39B White grains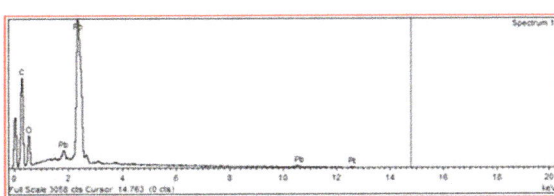

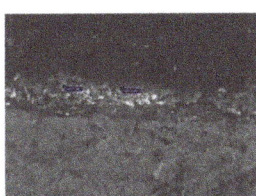 Sample 39B Blue grains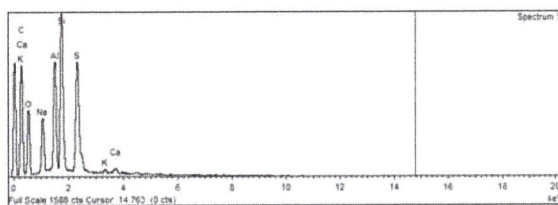

Figure 6. (**left**) SEM micrographs and (**right**) the corresponding EDS spectra of the SG_39B microsample cross-section.

3.5. SG_41A Microsample

This microsample is constituted by three layers:
- Plaster made of a light-colored binder (calcium carbonate), consisting of aggregates of variable grain size and color;
- White layer;
- Continuous, compact, and homogeneous purplish layer.

The EDS analysis, acquired along the dashed line in the SEM micrograph, showed the trends of the characteristic elements Ca, Fe, and a slightly higher amount of manganese (Mn), Figure 7. From the distribution of the elements along the cross-section of the sample, from the plaster toward the superficial pictorial layers, a high Ca content could be observed, attributable to the calcite present in the plaster and in the first white layer, followed by an intense peak of Fe in correspondence with the purplish layer, which, together with the presence of small quantities of Mn, suggests the use of umber (iron and manganese oxides). This hypothesis was confirmed by the XRF spectra because the same elements were identified. As for the previous microsamples, the IR analysis showed the presence of calcium carbonate, Mowilith DM 912, and whewellite and weddellite. The FT-IR bands at 2900, 2839, 1620, 1121, 1033, 940, and 876 cm^{-1} confirmed the presence of the pigment *caput mortuum* mixed with gypsum [20].

3.6. SG_42 Microsample

The microsample is constituted by four layers:
- Plaster made of a light-colored binder (calcium carbonate), consisting of aggregates of variable grain size and color;
- First continuous and homogeneous white layer (zinc white);
- Second continuous and homogeneous purplish-red layer (*caput mortuum*);
- Continuous and inhomogeneous blue layer. The presence of grains of variable size and of different colors, particularly green, can be observed.

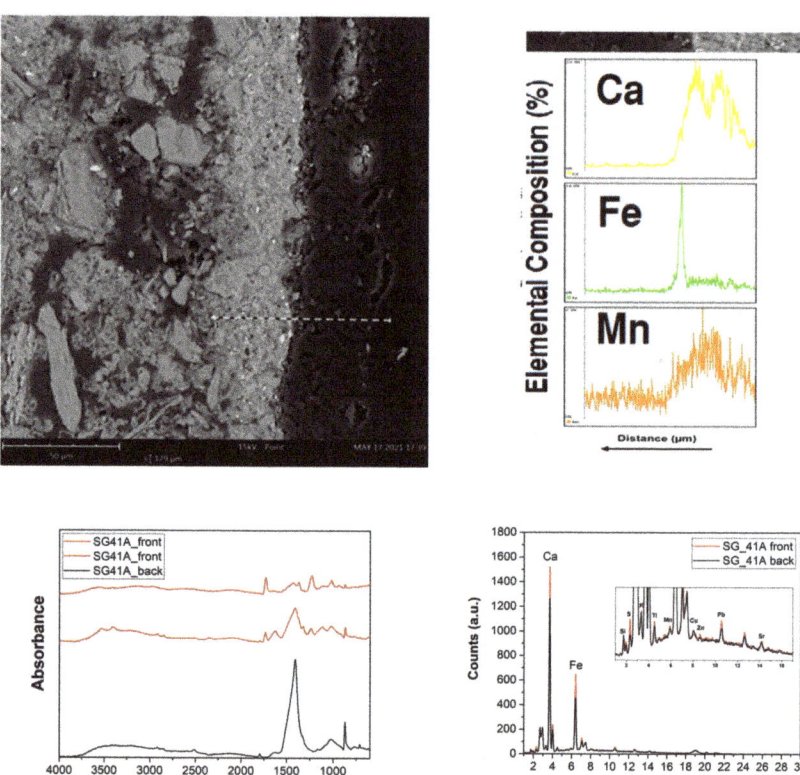

Figure 7. (**top left,right**) SEM micrograph and the corresponding EDS profile of the SG_41 microsample cross-section. The EDS investigation was carried out on the dashed line to evaluate the distribution of Ca, Fe, and Mn from back to front of the sample. (**bottom left**) FT-IR and (**bottom right**) XRF acquired on the front and back of the SG_41 microsample before preparation of the cross-section.

The EDS analysis, acquired along the dashed line reported in the SEM micrograph (Figure 8), showed the trends of the distribution of the characteristic elements Ca, zinc (Zn), Fe, and copper (Cu) along the cross-section of the sample from the plaster toward the surface paint layers. A high Ca content, attributable to the calcite of the plaster, was observed, followed by an increase in the Zn and Fe content, indicating the presence of zinc oxides in the white layer and of iron oxides in the red purplish layer, respectively, along with a subsequent increase in the Cu content in correspondence with the blue layer, attributable to the presence of a copper-based pigment such as azurite. Since the azurite is a water-soluble substance and reacts with the lime present in the fresh plaster, it was not applied a fresco, but a secco. On the other hand, the presence of green grains was evident under the optical microscope (Figure 9), which could be an indication of malachite, and an indication of the well-known reaction of conversion of azurite in malachite [21]. However, the results did not allow excluding the presence of other green–blue Cu-based pigments [22]. The barium (Ba) signals in the XRF spectra could indicate the presence of impurities in the copper-based pigment.

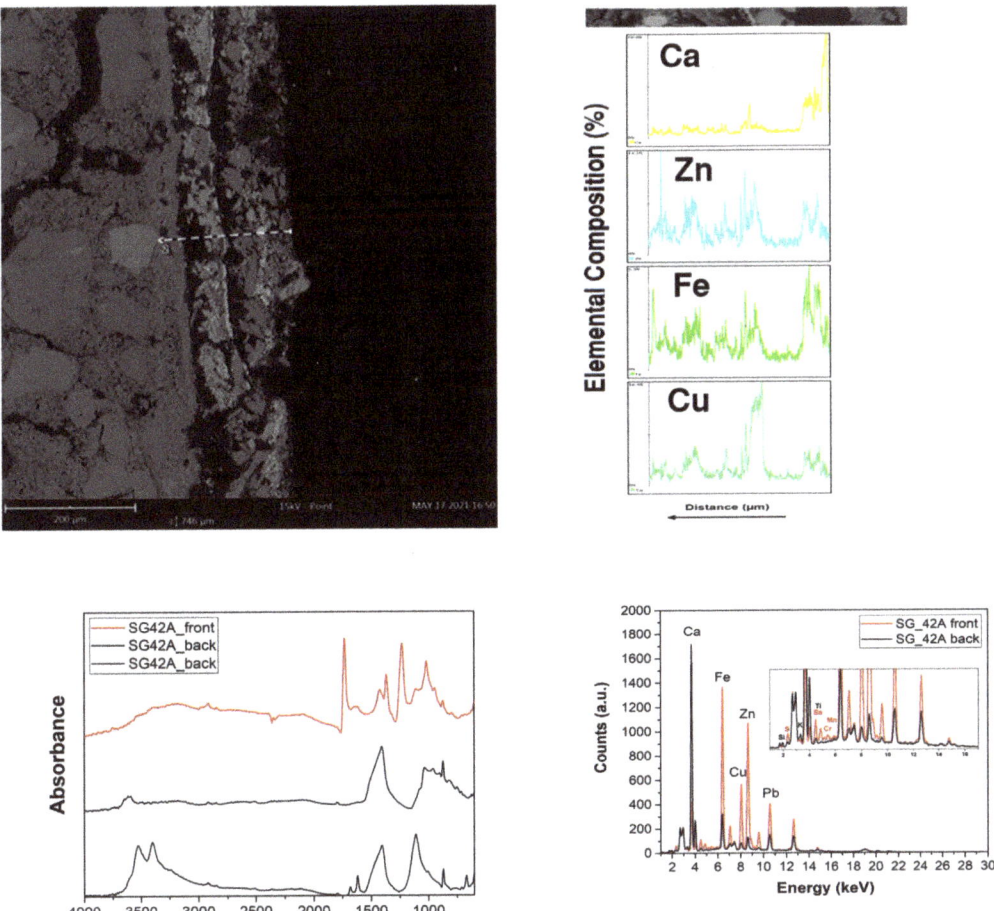

Figure 8. (**top left,right**) SEM micrograph and the corresponding EDS profile of the SG_42 microsample cross-section. The EDS investigation was performed on the dashed line. (**bottom left**) FT-IR and (**bottom right**) XRF spectra acquired on the front and back of the SG_42 microsample before preparation of the cross-section.

The presence of zinc oxide in the first white layer and of iron oxide added with manganese oxides in the second purplish-red layer was confirmed by the XRF spectra, suggesting the presence of the pigment *caput mortuum and umber* pigment.

As for the previously described microsamples, the FT-IR analysis showed the presence of calcium carbonate, Mowilith DM 912, and the pigment *caput mortuum* or morellone mixed with gypsum.

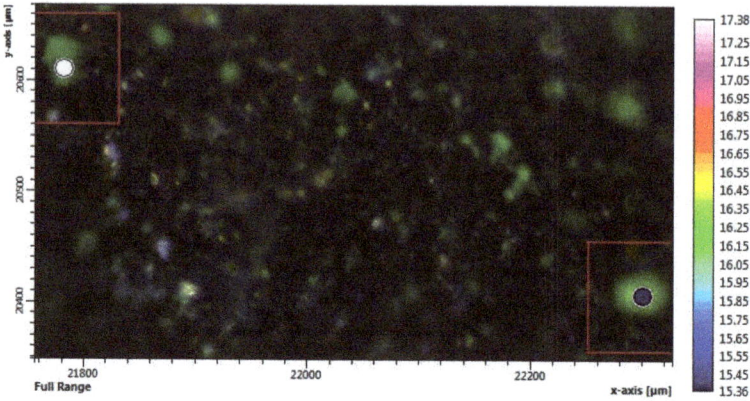

Figure 9. Image of the surface observed under the optical microscope.

4. Conclusions

The investigation here reported gave an overview of the pigments, the conservation state, the executive technique, and the trace of past retouching. For all samples, the plaster made of calcium carbonate ($CaCO_3$), consisting of aggregates of variable grain size and different colors, and the presence of a copolymer of vinyl-acrylate acetate/Mowilith DM912 [14], probably used in previous restoration work as protective, were identified. The presence of weddellite and whewellite is ascribable to some microbiological attack, and the presence of gypsum is an indication of the not good conservation state.

The painter used the traditional fresco technique with the white primer of calcium carbonate before applying the pigments.

In the vault, the following original pigments have been identified:

– Black pigment probably obtained from natural earth (black earth, a mixture of clay and calcium carbonate, iron and manganese)
– Dark gray: *caput mortuum or morellone* (mixture of organic black and ochre)
– Red pigment: red ochre/hematite
– White pigment: lead white
– Red–violet primer (*caput mortuum or morellone*)
– Green pigment: green earth
– Blue pigment: ultramarine blue
– The pigments, used during previous restoration work, are:
– White pigment: zinc white
– Red pigment (red ochre), applied a secco, retouching;
– Blue pigment (Prussian blue), applied a secco, retouching.

The blue pigment azurite applied a secco, partially converted into the green malachite, could be both original and applied by retouching.

The same pigments have been identified on other areas of the painting in the framework of the same campaign [23].

Through the recent campaign of investigations, representing the most complete study ever on the cycle of Saturnino Gatti in San Panfilo, we added some important details to the written information. From the beginning, the aim of carrying out scientific investigations was to find answers and acquire new information on the basis of objective data to elaborate the next restoration project. However, it cannot be excluded that future intervention on the paintings may still reveal some surprises, and that new details can be highlighted, as the restoration has always represented a privileged moment to increase knowledge of the work. Since the painting had some dry finishings (original), which unfortunately have been lost, the future restoration will also help in their identification and in their differentiation from the repaintings.

Author Contributions: M.L.S., conceptualization and methodology; M.F.F.M. historical research; V.C.C., M.L.S., A.P.S., E.P. and F.B., data curation and writing; V.C.C., µ-IR and OM; F.A., XRF and SEM investigations. A.P.S., E.P. and F.B., OM and SEM investigations; D.G., Raman investigation; R.C.P. and M.L.S., writing—review and editing. All authors have read and agreed to the published version of the manuscript.

Funding: This work was performed in the framework of the Project Torninparte "Archeometric investigation of the pictorial cycle of Saturnino Gatti in Tornimparte (AQ, Italy)" sponsored in 2021 by the Italian Association of Archeometry AIAR (www.associazioneaiar.com). F.A. thanks the MIUR for the Project PON Ricerca e Innovazione 2014–2020—Avviso DD 407/2018 "AIM Attrazione e Mobilità Internazionale" (AIM1808223).

Institutional Review Board Statement: Not applicable.

Informed Consent Statement: Not applicable.

Data Availability Statement: The datasets generated during and/or analyzed during the current study are available from the corresponding author on reasonable request. Samples are stored at the STEBICEF Department, University of Palermo (Italy).

Acknowledgments: The cross-sections were prepared by LabSTONE snc (http://www.labstone.it, accessed on 16 June 2023).

Conflicts of Interest: The authors declare no conflict of interest.

References

1. Galli, A.; Alberghina, M.F.; Re, A.; Magrini, D.; Grifa, C.; Ponterio, R.C.; La Russa, M.F. Special Issue: Results of the II National Research project of AIAr: Archaeometric study of the frescoes by Saturnino Gatti and workshop at the church of San Panfilo in Tornimparte (AQ, Italy). *Appl. Sci.* **2023**; under review.
2. Ricci, S. Tornimparte, a Mimesis of Florence in Abruzzo. New insights into Saturnino Gatti's art. *Appl. Sci.* **2023**; under review.
3. Arbace, L.; Di Paolo, G. *I Volti Dell'anima, Saturnino Gatti: Vita e Opere Di Un Artista Del Rinascimento*; Paolo De Siena Editore: Pescara, Italy, 2012; ISBN 8896341116.
4. Bonizzoni, L.; Caglio, S.; Galli, A.; Lanteri, L.; Pelosi, C. Materials and technique: The first look at Saturnino Gatti. *Appl. Sci.* **2023**, *13*, 6842. [CrossRef]
5. Bonizzoni, L.; Caglio, S.; Galli, A.; Germinario, C.; Izzo, F.; Magrini, D. Identifying original and restoration materials through spectroscopic analyses on Saturnino Gatti mural paintings: How far a non-invasive approach can go. *Appl. Sci.* **2023**, *13*, 6638. [CrossRef]
6. Andreotti, A.; Izzo, F.C.; Bonaduce, I. Archaeometric study of the mural paintings by Saturnino Gatti and workshop in the Church of San Panfilo-Tornimparte (AQ). The study of organic materials. *Appl. Sci.* **2023**; under review.
7. Defeyt, C.; Marechal, D.; Vandepitte, F.; Strivay, D. Rethinking Jacques-Louis David's Marat assassiné through material evidences. *Herit. Sci.* **2023**, *11*, 21. [CrossRef]
8. Dal Fovo, A.; Mattana, S.; Ramat, A.; Riitano, P.; Cicchi, R.; Fontana, R. Insights into the stratigraphy and palette of a painting by Pietro Lorenzetti through non-invasive methods. *J. Cult. Herit.* **2023**, *61*, 91–991. [CrossRef]
9. Mannetti, T.R.; Chelli, N.; Vecchioli, G. *Saturnino Gatti Nella Chiesa Di San Panfilo a Tornimparte*; L'Aquila Publishing: L'Aquila, Italy, 1992.
10. Saladino, M.L.; Ridolfi, S.; Carocci, I.; Chillura Martino, D.; Lombardo, R.; Spinella, A.; Traina, G.; Caponetti, E. A multi-analytical non-invasive and micro-invasive approach to canvas oil paintings. General considerations from a specific case. *Microchem. J.* **2017**, *133*, 607–613. [CrossRef]
11. Mollica Nardo, V.; Renda, V.; Parrotta, F.; Bonanno, S.; Anastasio, G.; Caponetti, E.; Saladino, M.L.; Vasi, C.S.; Ponterio, R.C. Non invasive investigation of the pigments of wall painting in S. Maria delle Palate di Tusa (Messina, Italy). *Heritage* **2019**, *2*, 2398–2407. [CrossRef]
12. Armetta, F.; Cardenas, J.; Caponetti, E.; Alduina, R.; Presentato, A.; Vecchioni, L.; di Stefano, P.; Spinella, A.; Saladino, M.L. Materials, conservation state and environmental monitoring of paintings in the Santa Margherita cliff cave. *Environ. Sci. Pollut. Res. J.* **2021**. [CrossRef]
13. Pouchou, J.L.; Pichoir, F. Quantitative Analysis of Homogeneous or Stratified Microvolumes Applying the Model "PAP". In *Electron Probe Quantitation*; Springer: Boston, MA, USA, 1991; pp. 31–75.
14. Horton-James, D.; Walston, S.; Zounis, S. Evaluation of the Stability, Appearance and Performance of Resins for the Adhesion of Flaking Paint on Ethnographic Objects. *Stud. Conserv.* **1991**, *36*, 203–221. [CrossRef]

15. Frost, R.L. Raman spectroscopy of natural oxalates. *Anal. Chim. Acta* **2004**, *517*, 207214. [CrossRef]
16. Henderson, E.J.; Helwig, K.; Read, S.; Rosendahl, S.M. Infrared chemical mapping of degradation products in cross-sections from paintings and painted objects. *Herit. Sci.* **2019**, *7*, 1–15. [CrossRef]
17. Wei, S.; Pintus, V.; Schreiner, M. Photochemical degradation study of polyvinyl acetate paints used in artworks by Py–GC/MS. *J. Anal. Appl. Pyrolysis* **2012**, *97*, 158–163. [CrossRef] [PubMed]
18. Guglielmi, V.; Andreoli, M.; Comite, V.; Baroni, A.; Fermo, P. The combined use of SEM-EDX, Raman, ATR-FTIR and visible reflectance techniques for the characterisation of Roman wall painting pigments from Monte d'Oro area (Rome): An insight into red, yellow and pink shades. *Environ. Sci. Pollut. Res.* **2022**, *29*, 29419–29437. [CrossRef] [PubMed]
19. Giannini, C. *Dizionario del Restauro. Tecniche, Diagnostica, Conservazione*; Nardini Editore: Firenze, Italy, 2010; pp. 113–114.
20. De Oliveira, L.F.; Edwards, H.G.; Frost, R.L.; Kloprogge, J.T.; Middleton, P.S. Caput mortuum: Spectroscopic and structural studies of an ancient pigment. *Analyst* **2002**, *127*, 536–541. [CrossRef] [PubMed]
21. Saladino, M.L.; Ridolfi, S.; Carocci, I.; Chirco, G.; Caramanna, S.; Caponetti, E. A multi-disciplinary investigation of the "Tavolette fuori posto" of the main hall wooden ceiling of the "Steri" (Palermo, Italy). *Microchem. J.* **2016**, *126*, 132–137. [CrossRef]
22. Zaffino, C.; Guglielmi, V.; Faraone, S.; Vinaccia, A.; Bruni, S. Exploiting external reflection FTIR spectroscopy for the in-situ identification of pigments and binders in illuminated manuscripts. Brochantite and posnjakite as a case study. *Spectrochim. Acta Part A Mol. Biomol. Spectrosc.* **2015**, *136*, 1076–1085. [CrossRef] [PubMed]
23. Briani, F.; Caridi, F.; Ferella, F.; Gueli A., M.; Marchegiani, F.; Nisi, S.; Paladini, G.; Pecchioni, E.; Politi, G.; Santo, A.P.; et al. Multi-Technique Characterization of Painting Drawings of the Pictorial Cycle at the San Panfilo Church in Tornimparte (AQ). *Appl. Sci.* **2023**, *13*, 6492. [CrossRef]

Disclaimer/Publisher's Note: The statements, opinions and data contained in all publications are solely those of the individual author(s) and contributor(s) and not of MDPI and/or the editor(s). MDPI and/or the editor(s) disclaim responsibility for any injury to people or property resulting from any ideas, methods, instructions or products referred to in the content.

Article

Multi-Technique Characterization of Painting Drawings of the Pictorial Cycle at the San Panfilo Church in Tornimparte (AQ)

Francesca Briani [1], Francesco Caridi [2], Francesco Ferella [3], Anna Maria Gueli [4], Francesca Marchegiani [3], Stefano Nisi [3], Giuseppe Paladini [2], Elena Pecchioni [5], Giuseppe Politi [4], Alba Patrizia Santo [5], Giuseppe Stella [4] and Valentina Venuti [2,*]

1. Adarte S.n.c., Via Agnoletti Enriques Annamaria 3, 50141 Firenze, Italy; francescabriani@adartesnc.it
2. Dipartimento di Scienze Matematiche e Informatiche, Scienze Fisiche e Scienze della Terra, Università degli Studi di Messina, V. le F. Stagno D'Alcontres 31, 98166 Messina, Italy; fcaridi@unime.it (F.C.); gpaladini@unime.it (G.P.)
3. CHNet INFN/Laboratori Nazionali del Gran Sasso, Via G. Acitelli 22, Assergi, 67100 L'Aquila, Italy; francesco.ferella@lngs.infn.it (F.F.); francesca.marchegiani@lngs.infn.it (F.M.); stefano.nisi@lngs.infn.it (S.N.)
4. Dipartimento di Fisica e Astronomia "Ettore Majorana", Università degli Studi di Catania, Via S. Sofia 64, 95123 Catania, Italy; anna.gueli@unict.it (A.M.G.); giuseppe.politi@ct.infn.it (G.P.); giuseppe.stella@dfa.unict.it (G.S.)
5. Dipartimento di Scienze della Terra, Università degli Studi di Firenze, Via La Pira 4, 50121 Firenze, Italy; elena.pecchioni@unifi.it (E.P.); alba.santo@unifi.it (A.P.S.)
* Correspondence: vvenuti@unime.it

Abstract: We present some results, obtained using a multi-scale approach, based on the employment of different and complementary techniques, i.e., Optical Microscopy (OM), Scanning Electron Microscopy-Energy Dispersive X-ray Spectroscopy (SEM-EDS), X-ray diffraction (XRD), Raman and μ-Raman spectroscopy, Fourier transform infrared (FT-IR) spectroscopy equipped with Attenuated Total Reflectance (ATR) analyses, Inductively Coupled Plasma–Mass Spectrometry (ICP-MS), and Thermal Ionization Mass Spectrometry (TIMS), of an integrated activity focused on the characterization of micro-fragments of original and previously restored paintings of the pictorial cycle at the San Panfilo Church in Tornimparte, sampled from specific areas of interest. The study was aimed, on one hand, at the identification of the overlapping restoration materials used during previous conservation interventions (documented and not), and, on the other hand, at understanding the degradation phenomena (current or previous) of the painted surfaces and the architectural structures. The study of stratigraphy allowed us to evaluate the number of layers and the materials (pigments, minerals, and varnishes) present in each layer. As the main result, the identification of blue, black, yellow, and red pigments (both ancient and modern) was achieved. In the case of blue pigments, original (azurite and lazurite) and retouching (Prussian blue and phthalo blue) materials were recognized, together with alteration products (malachite and atacamite). Traces of yellow ochre were found in the yellow areas, and carbon black in the blue and brown areas. In the latter, hematite and red ochre pigments were also recognized. The obtained results are crucial to support the methodological choices during the restoration intervention of the site, and help to ensure the compatibility principles of the materials on which a correct conservative approach is based.

Keywords: San Panfilo Church; multi-scale approach; frescoes; pigments; degradation phenomena; restoration

1. Introduction

The Italian Association of Archaeometry (AIAr) stipulated a Scientific Agreement with the Regional Secretariat of the Ministry of Cultural Heritage and Activities and Tourism for Abruzzo a few years ago. The initiative, launched in collaboration with the Superintendency of Archaeology, Fine Arts and Landscape for the City of L'Aquila and the

Citation: Briani, F.; Caridi, F.; Ferella, F.; Gueli, A.M.; Marchegiani, F.; Nisi, S.; Paladini, G.; Pecchioni, E.; Politi, G.; Santo, A.P.; et al. Multi-Technique Characterization of Painting Drawings of the Pictorial Cycle at the San Panfilo Church in Tornimparte (AQ). *Appl. Sci.* **2023**, *13*, 6492. https://doi.org/10.3390/app13116492

Academic Editor: Asterios Bakolas

Received: 28 April 2023
Revised: 22 May 2023
Accepted: 25 May 2023
Published: 26 May 2023

Copyright: © 2023 by the authors. Licensee MDPI, Basel, Switzerland. This article is an open access article distributed under the terms and conditions of the Creative Commons Attribution (CC BY) license (https://creativecommons.org/licenses/by/4.0/).

municipalities of the Crater, is aimed at the archaeometric study of an important cycle of frescoes in the Church of San Panfilo in Tornimparte (AQ), created along the vault and the perimeter walls of the apse by Saturnino Gatti (1491–1494), a pupil of Verrocchio, and his collaborators [1]. The cycle of frescoes depicts stories from the life and passion of Christ. In particular, in the middle of the vault, a colorful representation of Paradise is symbolized, with the Eternal Father surrounded by Angels and the Blessed. In the arch above the main altar, the Prophets, who foretold the coming of the Redeemer, are depicted, with the Archangel Gabriel in the act of announcing the birth of the Son of God to the Virgin on the sides. Around the apse, in five panels, the moments of the Redemption are reproduced upon a basement of painted marble, i.e.: *"Bacio di Giuda e la Cattura di Cristo"*, *"Flagellazione e l'Incoronazione di Spine"*, *"Crocifissione"*, *"Compianto sul Cristo morto"*, and *"Resurrezione"*.

The study intends to provide useful information from both a purely cognitive point of view, as a deepening of the artistic technique of the painter and his collaborators, and from a conservative point of view, in order to guide the methodological approaches for the restoration intervention of the site, helping restorers in applying a conservative approach that can guarantee reversibility and compatibility of the used materials [2–10].

The research project proposed by AIAr, by the title *"Studio archeometrico del ciclo pittorico di Saturnino Gatti e bottega presso la chiesa di San Panfilo in Villagrande di Tornimparte (AQ)"*, shared and approved with the stipulation of the aforementioned Scientific Agreement, foresees the achievement of the various objectives, including the compositional characterization of the original and restoration materials in view of the identification of the executive technique, and the documentation and understanding of the degradation phenomena, current or past, of the pictorial surfaces and the underlying architectural structures [11–16].

With this aim in mind, the present paper reports the results of a joint investigation focused on the multi-technique characterization of micro-fragments, taken from previously selected areas, using Optical Microscopy (OM), Scanning Electron Microscopy-Energy Dispersive X-ray Spectroscopy (SEM-EDS), X-ray diffraction (XRD), Raman and µ-Raman spectroscopy, Fourier transform infrared (FT-IR) spectroscopy equipped with Attenuated Total Reflectance (ATR) analyses, Inductively Coupled Plasma–Mass Spectrometry (ICP-MS), and Thermal Ionization Mass Spectrometry (TIMS).

It is worth remarking that, other than supporting the planning of the restoration intervention of the frescoes, these results will contribute to the promotion of this precious cultural asset, still almost unknown from an archaeometric point of view.

This paper contributes to the Special Issue "Results of the II National Research project of AIAr: archaeometric study of the frescoes by Saturnino Gatti and workshop at the church of San Panfilo in Tornimparte (AQ, Italy)" in which the scientific results of the II National Research Project conducted by members of the Italian Association of Archaeometry (AIAr) are discussed and collected.

For in-depth details on the aims of the project, see the introduction of the Special Issue [17].

2. Materials and Methods

2.1. Materials

A total of 15 micro-fragments were investigated, whose description and methods of analysis are reported in Table 1, taken from different sampling areas, namely Panel A *"Il bacio di Giuda e la Cattura di Cristo"*, Panel D *"Compianto sul Cristo morto"*, and Panel E *"Resurrezione"* (see Figure 1).

Table 1. List of investigated samples, together with the approximate size, sampling area, description, and methods of analysis.

Sample	Size (cm)	Sampling Area	Description	Methods of Analysis
SG_18B	0.5	Panel A	Yellow-orange pictorial layer, whitish preparation/priming sample taken along an existing gap	Raman
SG_19	0.5	Panel A	*Detail of a leaf*—Original area (green pictorial layer on a brown-red layer taken up to white plaster support)	OM, SEM-EDS, XRD
SG_20	0.5	Panel A	Original area (blue pictorial layer on a brown-red layer + fragments of plaster below)	OM, SEM-EDS, XRD
SG_21A	0.6	Panel A	*Blue pattern above the window*—Original area (blue pictorial layer on a brown-red layer + fragments of plaster below)	OM, SEM-EDS, XRD
SG_22	0.6	Panel A	*Sky, green pattern*—Original area (green pictorial layer, due to degradation of an originally light-blue layer, on a red-brown layer (fragments of underlying plaster)	OM, SEM-EDS, XRD
SG_23B	0.6	Panel A	*Sky*—Area with drops and probable alterations (light-blue pictorial layer)	µ-Raman, FT-IR
SG_24	0.5	Panel A	*Landscape, area contiguous to the window*—Brown-green pictorial layer + white preparation/priming taken along a gap	Raman
SG_31	0.5	Panel A	Blue pictorial layer + white preparation/priming and sampling carried out in correspondence with a gap	Raman
SG_25B	0.6	Panel D	*Sky, upper portion*—Blue pictorial layer on a brown-red layer + fragments of plaster below	µ-Raman, FT-IR
SG_27B	0.5	Panel D	Light-blue/yellow pictorial layer + white preparation/priming	Raman
SG_28B	0.6	Panel D	Dark-green pictorial layer + plaster fragments. The area also features glazed pictorial additions	µ-Raman, FT-IR
SG_33	0.6	Panel E	Purple pictorial layer (area affected by protective agents that make the surface shiny with a "wax" effect) + plaster fragments	µ-Raman, FT-IR
SG_45B	0.5	Panel E	*Detail of the drapery of the loincloth of the risen Christ*—White pictorial layer applied on an underlying pictorial surface	ICP-MS, TIMS

2.2. Methods

2.2.1. OM Measurements

The stratigraphic study was carried out on polished cross sections of samples using a ZEISS Axio Scope A1 microscope (Carl Zeiss, Jena, Germany), equipped with a video camera, resolution of 5 megapixels, and image analysis software AxioVision (V1). The analyses were carried out in reflected light (Reflected Light Microscopy, RLM).

2.2.2. SEM-EDS Measurements

The morphological and semi-quantitative microchemical analyses were performed using a SEM-EDS electronic microscope (ZEISS EVO MA 15) (Carl Zeiss, Jena, Germany) with a W-filament equipped with an analysis system in energy dispersion EDS/SDD, Oxford Ultimax 40 (40 mm² with resolution 127 eV @5.9 keV) (Oxford Instruments, Abingdon,

UK) and the Aztec 5.0 SP1 software. The measurements were performed on carbon metalized cross sections of the samples under the following operating conditions: accelerating potential of 15 kV, 500 pA beam current, working distance between 9 and 8.5 mm; 20 s live time as an acquisition rate useful for archiving at least 600,000 cts, on co-standard, and process time 4 for point analyses; 500 µs pixel dwell time for recording 1024 × 768 pixel resolution maps. The software for the microanalysis was Aztec 5.0 SP1 software, which uses the XPP matrix correction scheme developed by Pouchou and Pichoir in 1991 [18]. This is a Phi-Rho-Z approach that employs exponentials to describe the shape of the $\varphi(\rho z)$ curve. The XPP matrix correction was chosen because it works well in situations with strong absorption, such as the analysis of light elements in a heavy matrix. The method is a "standardless" quantitative analysis, using pre-purchased standard materials for the calculations. The monitoring of constant analytical conditions (i.e., filament emission) is archived with repeated analyses of a Co-metal standard.

Figure 1. Sampling areas for the investigated micro-fragments.

2.2.3. XRD Measurements

The semi-quantitative mineralogical analyses of the bulk samples were performed using XRD with a Philips PW 1050/37 diffractometer (Philips, Almelo, The Netherlands) and a Philips X'Pert PRO data acquisition and analysis system operating at 40 kV–20 mA, with a Cu anode, a graphite monochromator, and a goniometry speed of 2°/min in a scan range between 5 and 70° θ; the slits are 1-01-1 and the detection limit is 4%.

2.2.4. Raman and µ-Raman Measurements

The raman measurements were performed using a portable Madatech Raman, with a wavelength of 785 nm, a max laser power of 350 mW, an optical-fiber-connected probe head in backscattering geometry, and a Peltier cooled silicon CCD detector. The operative window was 60–3200 cm^{-1} with a resolution of 4 cm^{-1}; the typical acquisition power and time were 30 mW and 30 s, respectively, on a spot of about 2 mm.

The μ-Raman measurements were performed using a portable Raman 'BTR 111 Mini RamTM' (Be&W TEK Inc., Newark, NJ, USA) spectrometer, by using an incident wavelength of 785 nm (diode laser), a max laser power of 280 mW, and a Thermoelectic (CE) Cooled 2048-pixel CCD detector. The 62–3153 cm^{-1} spectral range was investigated, with a resolution of 10 cm^{-1}, an acquisition power of 30 mW and an acquisition time of 10 s × 32 scans. The system was equipped with a BAC151B Raman microscope. The 80× objective, with a working distance of 1.25 mm and a laser beam spot size of 26 μm.

The identification of the peaks was obtained by comparing the experimental spectra with those reported in various databases and the literature [19–22]

2.2.5. FT-IR Measurements

The FT-IR measurements were conducted using a Perkin Elmer Spectrum 100 spectrophotometer, in Attenuated Total Reflectance (ATR) mode, directly on the fragment under examination, without any a priori preparation. The spectra were recorded at a resolution of 4 cm^{-1} between a 500 and 4000 cm^{-1} wavenumber range.

The identification of the peaks was achieved by comparing the experimental spectra with those reported in various databases and in the literature [23,24].

2.2.6. ICP-MS Measurements

The elemental characterization and lead isotope measurement of the sample SG_45B, which was supposed to be "lead white", were performed at the mass spectrometer facility at CHNet-LNGS [25] to study the provenance of the raw material used to produce the pigment. The Inductively Coupled Plasma–Mass Spectrometer (ICP-MS) model 7500a by Agilent Technologies was used for the qualitative elemental characterization. About 5 mg of the sample were dissolved in 5% solution of the nitric acid and properly diluted before instrumental analysis. The concentrations were determined in Semi-quantitative mode calibrating the instrumental response based on a single reference solution containing 10 ng·g^{-1} of Li, Y, Ce, and Tl in order to cover the whole mass range. The uncertainties achieved operating in this mode are about 20% of the concentration value.

2.2.7. TIMS Measurements

The Thermal Ionization Mass Spectrometry (TIMS) is a suitable technique to measure the lead isotope ratio of the sample. The multi-collector spectrometer Finnigan MAT 262 with hardware and software upgraded to TIBox by Spectromat-GmbH (Bremen, Germany) was used [26]. The concentration of Pb measured in the sample using ICP-MS was about 450 mg·kg^{-1}, so its extraction and purification using selective chromatographic resin supplied by Triskem (Bruz, France) was mandatory. Next, 10 mL of the solution and 0.1 M of ammonium oxalate, used to elute the lead from the resin, were evaporated and treated in the oven at 300 °C for 3 h to remove the organic residual. Finally, the lead was recovered using 25 μL of 1% nitric acid solution and then 5 μL were loaded on the "zone-refined" rhenium filament for TIMS measurement.

3. Results and Discussion

3.1. Panel A "Il Bacio di Giuda e la Cattura di Cristo"

In the case of the SG_18B sample, the spectrum obtained from the yellow-orange area (Figure 2) presents two structures, at ~1180 and ~1270 cm^{-1}, that could be attributed to the goethite of yellow ochre; the contribution of this mineral expected at ~440 cm^{-1} is probably masked by the high value of fluorescence, mainly due to the binding media. Additionally, the contribution of calcite, reasonably coming from the preparation layer, is visible.

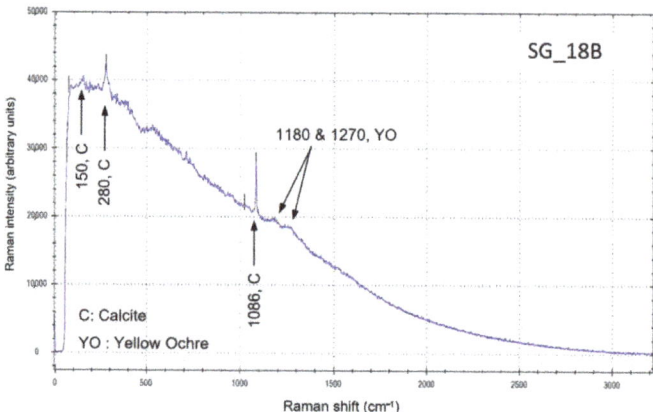

Figure 2. Raman spectrum collected on yellow-orange area of SG_18B sample.

The sample SG_19, collected from a green leaf (Figure 1), shows a four-layer stratigraphy (Figure 3): starting from the surface, we can observe a greenish layer (A) with a thickness of about 25 µm, a blue layer (B), with the same thickness, related to the sky, a red layer (C), and the substrate consisting of the mortar (D). The elemental analysis (SEM-EDS; Table 2) shows the presence of abundant copper in the green and blue grains of the A and B layers, which is due to the use of malachite and azurite pigments. The presence of large amounts of calcium in layers A and B could indicate the use of a white pigment (*Bianco di San Giovanni*?) to lighten the green and blue tones; layer C is made of red ochre film (*morellone*) to protect the pigments from the alkaline effects of the mortar. The XRD analysis of the selected green grains from this sample shows the presence of malachite, along with traces of quartz (Table 3).

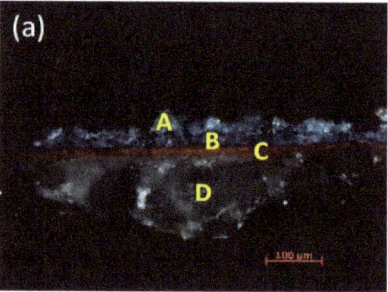
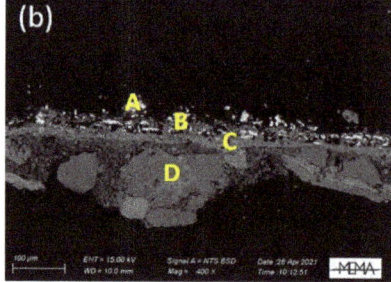

Figure 3. (**a**) Micrographs of SG_19 cross section obtained using OM, showing the four-layer stratigraphy; (**b**) SEM-EDS back-scattering electron image of the four layers.

The sequence of layers of the sample SG_20 (obtained from the base coat of the sky, see Figure 1) consists of only three layers (Figure 4): a very thin (less than 20 µm) blue film (A), a red layer (probably *morellone*) (B), and the mortar substrate (C). The SEM-EDS (Table 2) investigation displays the use of azurite for the outer layer, taking into account the copper content, and red ochre (*morellone*) for the second one. As well as in the sample SG_19, a similar consideration about the presence of a high percentage of calcium in the blue layer is necessary.

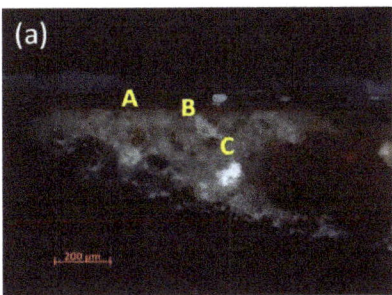
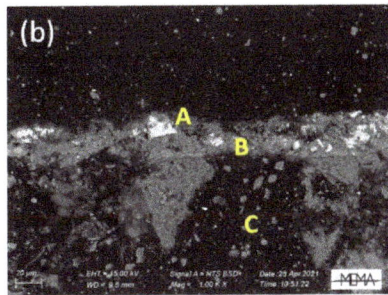

Figure 4. (**a**) Micrographs of SG_20 cross section obtained using OM, showing the three-layer stratigraphy; (**b**) SEM-EDS back-scattering electron image of the three layers.

As for the fragment SG_21A, taken from a blue wash above the window (Figure 1), the observation under the light microscope shows two different shades of blue on the surface (layers A and B), overlying a red layer (C), and the substrate of mortar (D) (Figure 5). The blue layers differ in tonality. The first one (A) is lighter and contains black grains of small size, while the second (B) displays a more intense blue tone and a coarser grain size. In addition, the B layer is not continuous along the section, suggesting a loss of colour from the original film in some areas, and supporting the presence of a secondary integration with the lighter blue pigment belonging to layer A. The SEM-EDS analysis (Table 2) of layer A detected the presence of iron and zinc, suggesting the use of a mixture of Prussian blue and Zinc white. These materials clearly attest to the non-original nature of layer A, allowing us to date the intervention to not before the second half of the XVIII century. The punctual investigation of the black grains observed inside layer A revealed the presence of calcium and phosphorus, due to the use of a bone black pigment, probably added to make the tone of blue darker (too much light), making it uniform with coat B. The elemental analysis confirms the use of azurite for the coarse blue grains of layer B and red ochre for the *morellone* C layer.

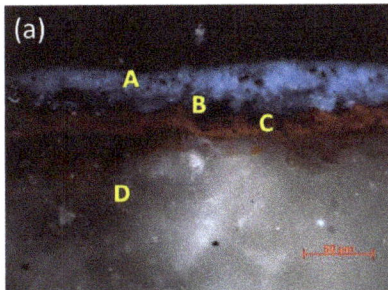
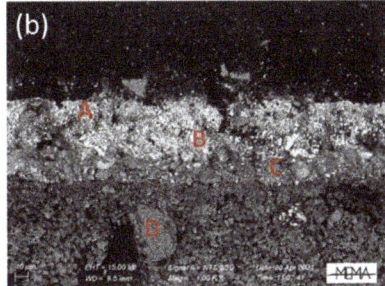

Figure 5. (**a**) Micrographs of SG_21A cross section obtained using OM, showing the four-layer stratigraphy; (**b**) SEM-EDS back-scattering electron image showing the four layers.

The SG_22 greenish fragment was collected from the sky background (Figure 1) that was probably subjected to chromatic alteration. The cross section, observed under the optical microscope, features three layers, including an altered green layer on the surface (A), a thin red layer, probably *morellone* (B), and a substrate consisting of mortar (C) (Figure 6). The SEM-EDS elemental analysis (Table 2) confirms the hypothesis of colour shading in the surface pigment; indeed, copper and chlorine were detected on the A layer, in relation to the formation of atacamite, a copper chloride ($Cu_2Cl(OH)_3$) and a degradation product of azurite. The presence of iron in the red layer attests to the use of the *morellone* (B).

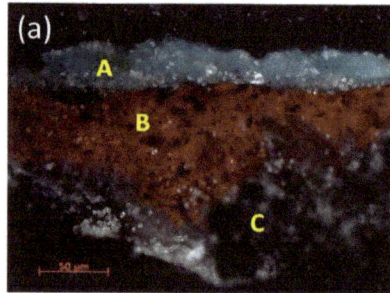 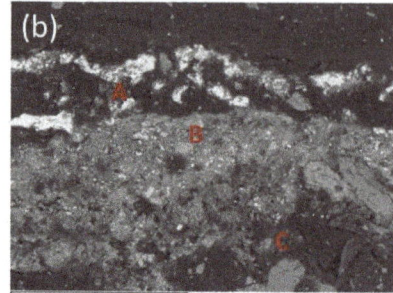

Figure 6. (a) Micrographs of SG_22 cross section obtained using OM, showing the three-layer stratigraphy; (b) SEM-EDS back-scattering electron image of the three layers.

Table 2. EDS analysis (weight %) of samples SG_19, SG_20, SG_21A, and SG_22 cross sections.

Sample ID	P_2O_5	MgO	Al_2O_3	SiO_2	SO_3	Cl	K_2O	CaO	FeO	CuO	ZnO
SG_19 Greenish layer	-	3.58	1.89	12.33	3.83	-	-	17.75	3.02	55.54	-
SG_19 Blue layer	-	1.84	1.29	13.27	1.96	-	1.38	41.76	3.24	34.50	-
SG_20 Blue layer	-	1.54	-	10.99	2.14	-	0.90	45.24	-	38.48	-
SG_21A A blue layer	-	-	2.01	20.10	3.44	-	-	5.70	2.06	7.49	58.43
SG_21A A layer, black grains	29.93	-	-	-	3.04	-	-	36.19	-	-	30.13
SG_21A B layer, blue grains	-	-	-	-	-	-	-	1.06	-	98.94	-
SG_22 Greenish layer	-	1.86	1.68	4.77	-	24.73	0.63	4.95	-	61.38	-

A supplementary fragment of the sample SG_22 was investigated using XRD (Table 3). The analysis reveals and confirms the presence of atacamite, resulting from azurite alteration under specific conditions; the presence of quartz, calcite, plagioclase, mica, and chlorite are to be attributed to the composition of the mortar that polluted the analysis of the pigment.

Table 3. XRD semi-quantitative results: XXX = high content; XX = medium content; X = low content; tr = traces.

Sample ID	Quartz	Calcite	Plagioclase	Mica	Malachite	Chlorite	Atacamite
SG_19	tr	-	-	-	XXX	-	-
SG_22	XX	X	X	tr	-	tr	tr

In the case of the SG_23B micro-fragment, the μ-Raman spectrum (Figure 7) collected on a blue area shows the presence of lazurite, as a blue pigment, mixed with carbon black (smoke black), calcite, and gypsum. Importantly, considering the specific historical–geographical context and execution technique of the whole pictorial cycle, the presence of gypsum can be considered as intentionally added by the painter as binding material for the pigments. However, the existence of gypsum, resulting from environmentally activated alteration processes that took place over time onto the topmost pictorial surface, cannot be excluded.

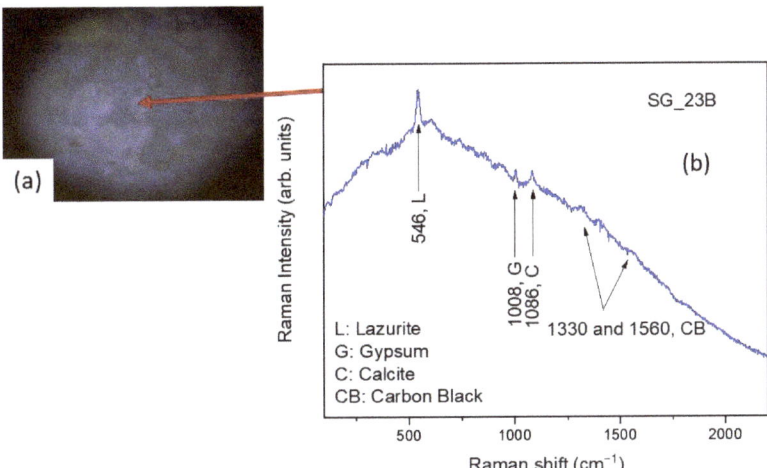

Figure 7. (**a**) Micro-photo of the analyzed blue area, (**b**) µ-Raman spectrum collected on the blue area of SG_23B sample.

The FT-IR spectrum of this sample (Figure 8) clearly highlighted the presence of the characteristic bands of calcite and gypsum. The band between ~890 cm^{-1} and ~1230 cm^{-1} is associated with silicates. No organic binders are observed.

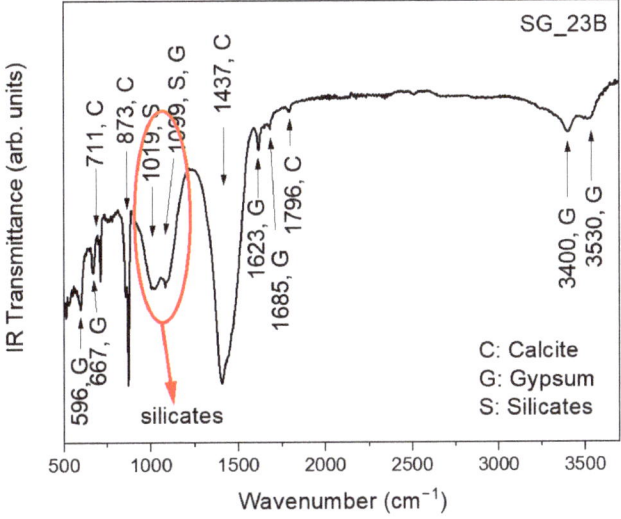

Figure 8. FT-IR spectrum of SG_23B sample.

This experimental evidence is in agreement with the use of the "*a fresco*" technique for the application of the pigment.

The spectrum (Figure 9) measured on the brown-green pictorial layer of SG_24 sample shows the presence of carbon black, probably used to darken the used pigment, that does not show any characteristic peak.

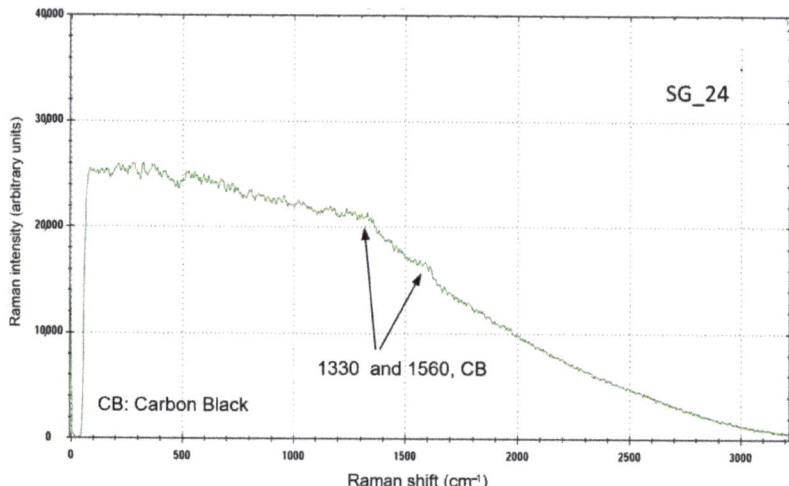

Figure 9. Raman spectrum collected on SG_24 brown-green sample.

The SG_31 blue sample shows a Raman spectrum (Figure 10) with the characteristic peak of calcite and those of carbon black. No marks of a blue pigment are visible.

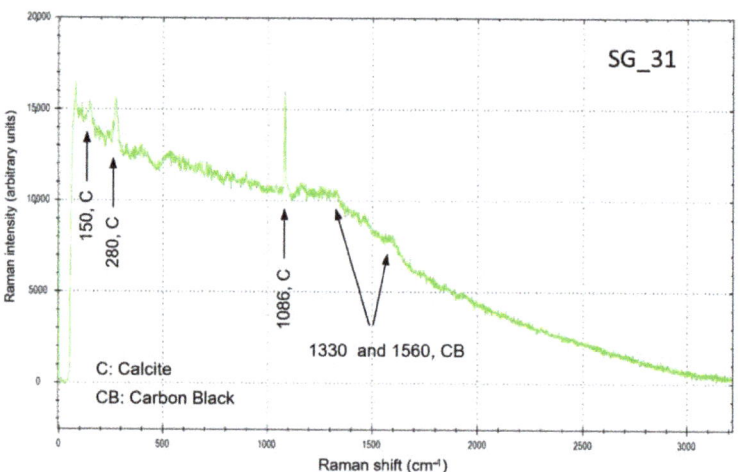

Figure 10. Raman spectrum collected on SG_31 blue sample.

3.2. Panel D "Compianto sul Cristo Morto"

First, a Raman analysis was carried out on a red area (Figure 11a) of the SG_25B microfragment, whose spectrum highlighted the presence of calcite and gypsum (Figure 11b). No bands attributable to the pigment were observed. Furthermore, a Raman investigation was carried out on a blue area (Figure 11c), whose spectrum (Figure 11d) allowed for highlighting the presence of lazurite [27,28], as an azure pigment, mixed with phthalo blue (synthetic inorganic pigment), in addition to the peak at 1086 cm^{-1} attributable to the calcite plaster.

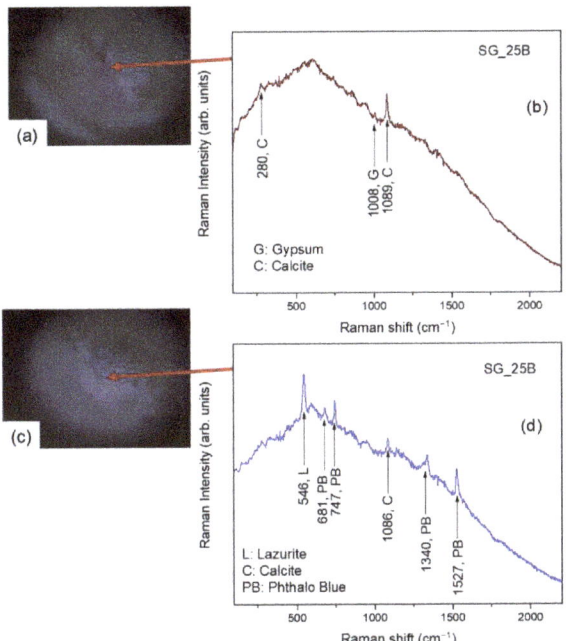

Figure 11. (**a**) Micro-photo of the analyzed red area, (**b**) μ-Raman spectrum collected on the red area of SG_25B sample, (**c**) micro-photo of the analyzed blue area, (**d**) μ-Raman spectrum collected on the blue area of SG_25B sample.

The FT-IR spectrum of the SG_25B sample, reported in Figure 12, appears entirely dominated by the vibrational contributions of calcite and gypsum. No organic binders are observed.

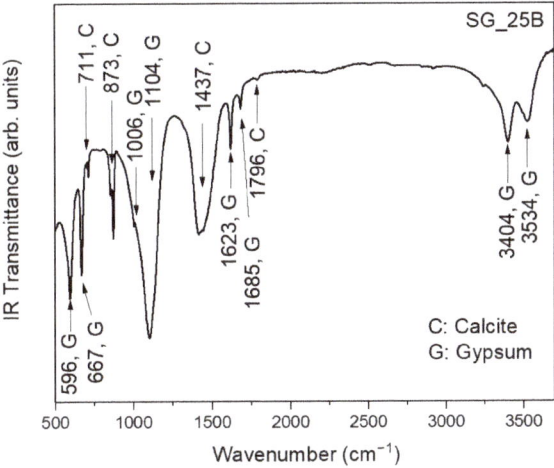

Figure 12. FT-IR spectrum of SG_25B sample.

In the light-blue/yellow area of the SG_27B sample (Figure 13), the Raman spectrum presents the peaks of calcite, and, again, the two structures of yellow ochre. No marks of a blue pigment, expected with 785 nm excitation laser in the region of 400–600 cm^{-1} for lazurite or azurite, are visible, probably masked by the high fluorescence.

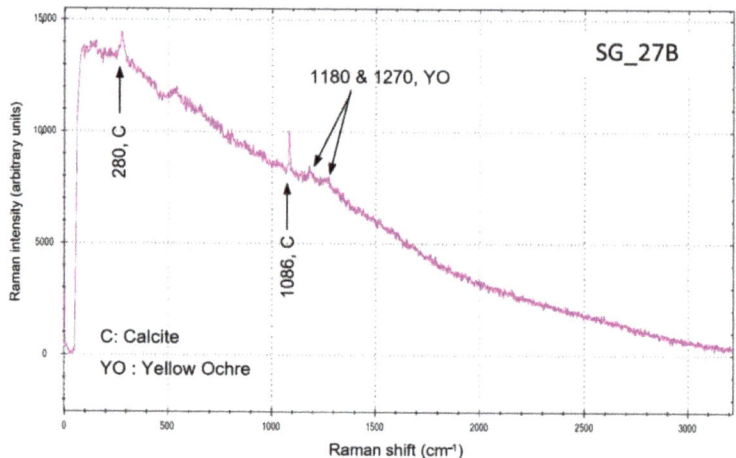

Figure 13. Raman spectrum collected on SG_27B light-blue/-yellow sample.

As regards the SG_28B sample, the µ-Raman observation highlights blue areas (Figure 14a). The spectrum (Figure 14b) exhibits all the characteristic peaks of phthalo blue.

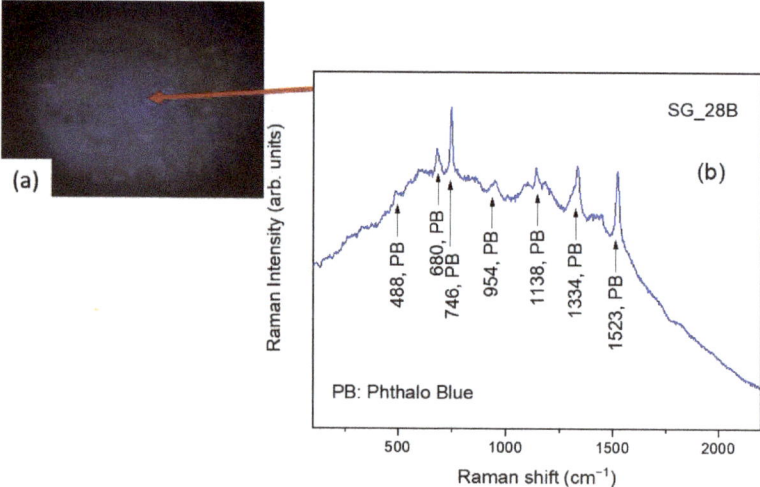

Figure 14. (a) Micro-photo of the analyzed blue area, (b) µ-Raman spectrum collected on the blue area of SG_28B sample.

The FT-IR spectrum of this sample (Figure 15), in addition to the usual contributions from calcite, highlighted the presence of the broadband between ~890 cm^{-1} and ~1230 cm^{-1} associated with the presence of silicates, together with the contributions at ~1633 cm^{-1} (HOH bending) and the broad band centered at ~3300 cm^{-1} (OH stretching), which is indicative of the presence of clayey minerals. No organic binders are observed.

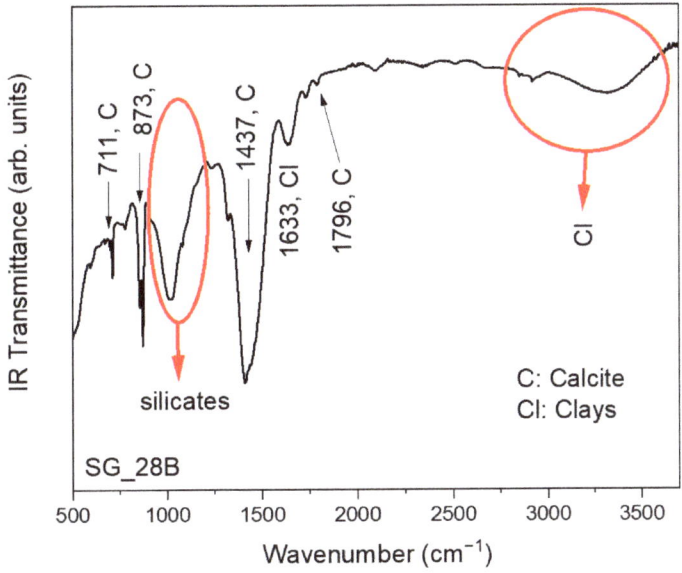

Figure 15. FT-IR spectrum of SG_28B sample.

3.3. Panel E "Resurrezione"

As far as the SG_33 micro-fragment is concerned, the brown area analyzed using µ-Raman (Figure 16) allowed for highlighting all the characteristic peaks of hematite, as a red pigment, mixed with carbon black, as well as calcite from the base.

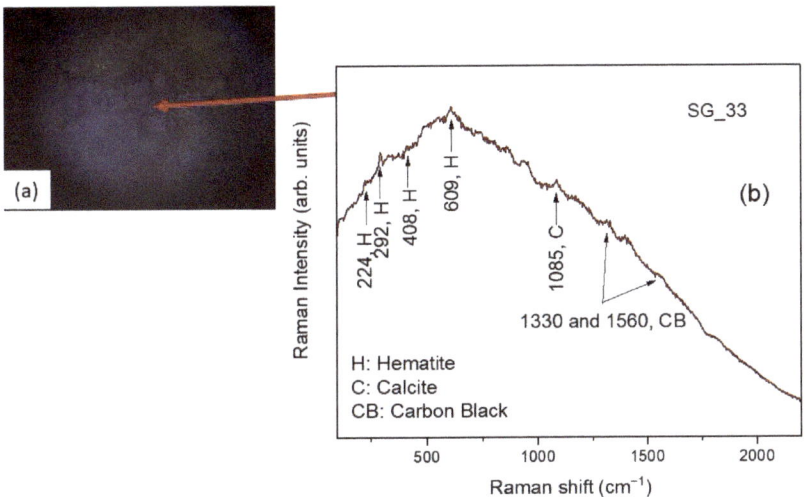

Figure 16. (**a**) Micro-photo of the analyzed brown area, (**b**) µ-Raman spectrum collected on the brown area of SG_33 sample.

The FT-IR spectrum (Figure 17) clearly revealed the presence of the typical vibrational bands of calcite.

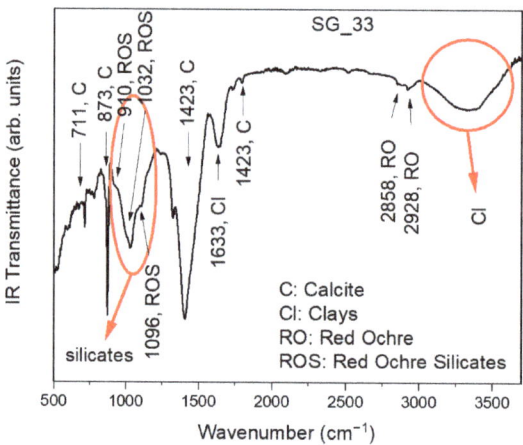

Figure 17. FT-IR spectrum of SG_33 sample.

Again, the broad band between ~890 cm^{-1} and ~1230 cm^{-1} associated with the presence of silicates appears evident. Within it, the peaks at ~910 cm^{-1}, ~1032 cm^{-1}, and ~1096 cm^{-1}, which can be associated to silicates, probably present in red ocher [29], can be distinguished. Finally, the band at ~1633 cm^{-1} and the broad band centered at ~3300 cm^{-1}, attributed to the OH groups, are indicative of the presence of clayey minerals. No organic binders are observed.

The elemental composition of the sample SG_45B was investigated using ICP-MS. The concentrations of the main detected elements are reported in Table 4.

Table 4. Elemental composition of the sample measured using ICP-MS. The uncertainties are 20% of the concentration values.

Sample ID	Element	Concentration [mg·kg^{-1}]	Concentration [%]
SG_45B	Na	265	0.03
	Mg	2100	0.2
	Ca	62,000	6.2
	Fe	1000	0.1
	Cu	74	0.007
	Zn	520	0.05
	Ba	1000	0.1
	Pb	450	0.045

The Pb contained in the sample constitutes a very small mass fraction (0.045%), so the analyzed material cannot be correctly identified using the "lead white" pigment, commonly used in these studies, which contains about 80% of lead. Even if the results allow for a partial identification of the sample, the most abundant compounds seem to be calcium based, such as gypsum or calcite.

The doubtful origin of the lead poses a basic uncertainty to study the provenance of the raw material used in the pictorial layer. Moreover, the availability of a single sample introduces uncertainty about the overall representativeness of the entire painting.

The lead isotope composition measured in the sample is quite rare due to the relatively high value of Pb207/206 and Pb208/206 ratio. Table 5 compares the isotopic ratios measured for SG_45B and those for some mining sites in the South of France that have the closest lead isotope signature related to the sample.

Table 5. Comparison between the lead isotope ratio of SG_45B sample with closest lead mines present in the database.

Sample	$^{207}Pb/^{206}Pb$	$^{208}Pb/^{206}Pb$	$^{206}Pb/^{204}Pb$	$^{207}Pb/^{204}Pb$	$^{208}Pb/^{204}Pb$
SG45B	0.8529 ± 0.0003	2.0900 ± 0.0007	18.26 ± 0.03	15.57 ± 0.03	38.16 ± 0.07
Férols, Montgaillard (France)	0.8525	2.0903	18.33	15.63	38.32
Lastours, Montagne Noire (France)	0.8535	2.0910	18.36	15.67	38.39
Cevennes, Massif Central (France)	0.8519	2.0915	18.36	15.64	38.40

In particular, the Férols mine, located in the region of present-day New Aquitaine, shows reasonable compatibility considering the limits discussed above.

4. Conclusions

In the present study, a multi-scale approach involving Optical Microscopy (OM), Scanning Electron Microscopy-Energy Dispersive X-ray Spectroscopy (SEM-EDS), X-ray diffraction (XRD), Raman and μ-Raman spectroscopy, Fourier transform infrared (FT-IR) spectroscopy equipped with Attenuated Total Reflectance (ATR) analyses, Inductively Coupled Plasma–Mass Spectrometry (ICP-MS), and Thermal Ionization Mass Spectrometry (TIMS) was employed in order to characterize, in the framework of the A.I.Ar. Research project *"Studio archeometrico del ciclo pittorico di Saturnino Gatti e bottega presso la chiesa di San Panfilo in Villagrande di Tornimparte (AQ)"*, the micro-fragments of the paintings of the pictorial cycle at the San Panfilo Church in Tornimparte, sampled from specific areas of interest.

Such a combined approach allowed us to successfully characterize, both at the elemental and molecular scales, the composition of the materials used by the artist in terms of preparatory components and pigmenting agents, furnishing novel insights into the execution technique of the master, the color palette, and the occurrence of nondocumented restoration treatments. Moreover, the analysis of the sample stratigraphy permitted a proper evaluation of the number of layers, together with the materials present in each of them. In this context, the measurements allowed us to confirm, among other aspects, the presence of both ancient and modern pigments, together with alteration products onto the pictorial surfaces of the cycle of frescoes.

It is worth noting that the whole set of results, obtained by the employment of a multi-scale approach involving different spatial regimes, represents an essential pre-requisite in view of the optimization of the latest restoration procedures and for the accurate selection, by restorers and conservators, of the protection and best-cleaning strategies to be applied.

Author Contributions: Conceptualization, F.B., F.C., F.F., A.M.G., F.M., S.N., G.P. (Giuseppe Paladini), E.P., G.P. (Giuseppe Politi), A.P.S., G.S. and V.V.; methodology, F.B., F.C., F.F., A.M.G., F.M., S.N., G.P. (Giuseppe Paladini), E.P., G.P. (Giuseppe Politi), A.P.S., G.S. and V.V.; validation, V.V.; formal analysis, F.B., F.C., F.F., A.M.G., F.M., S.N., G.P. (Giuseppe Paladini), E.P., G.P. (Giuseppe Politi), A.P.S., G.S. and V.V.; investigation, F.B., F.C., F.F., A.M.G., F.M., S.N., G.P. (Giuseppe Paladini), E.P., G.P. (Giuseppe Politi), A.P.S., G.S. and V.V.; data curation, F.B., F.C., F.F., A.M.G., F.M., S.N., G.P. (Giuseppe Paladini), E.P., G.P. (Giuseppe Politi), A.P.S., G.S. and V.V.; writing—original draft preparation, F.B., F.C., F.F., A.M.G., F.M., S.N., G.P. (Giuseppe Paladini), E.P., G.P. (Giuseppe Politi), A.P.S., G.S. and V.V.; writing—review and editing, F.B., F.C., F.F., A.M.G., F.M., S.N., G.P. (Giuseppe Paladini), E.P., G.P. (Giuseppe Politi), A.P.S., G.S. and V.V.; visualization, F.B., F.C., F.F., A.M.G., F.M., S.N., G.P. (Giuseppe Paladini), E.P., G.P. (Giuseppe Politi), A.P.S., G.S. and V.V.; supervision, V.V. All authors have read and agreed to the published version of the manuscript.

Funding: This work was performed in the framework of the Project Tornimparte—"Archeometric investigation of the pictorial cycle of Saturnino Gatti in Tornimparte (AQ, Italy)" sponsored in 2021 by the Italian Association of Archeometry AIAR (www.associazioneaiar.com).

Institutional Review Board Statement: Not applicable.

Informed Consent Statement: Not applicable.

Data Availability Statement: The data presented in this study are available on request from the corresponding author.

Conflicts of Interest: The authors declare no conflict of interest.

References

1. Latini, M. Chiesa di San Panfilo—Villagrande di Tornimparte (AQ). In *Guida alle Chiese d'Abruzzo*; Carsa Edizioni: Pescara, Italy, 2016; pp. 139–140. ISBN 978-88-501-0354-6.
2. Zuena, M.; Buemi, L.P.; Stringari, L.; Legnaioli, S.; Lorenzetti, G.; Palleschi, V.; Nodari, L.; Tomasin, P. An integrated diagnostic approach to Max Ernst's painting materials in his Attirement of the Bride. *J. Cult. Herit.* **2020**, *43*, 329–337. [CrossRef]
3. Venuti, V.; Fazzari, B.; Crupi, V.; Majolino, D.; Paladini, G.; Morabito, G.; Certo, G.; Lamberto, S.; Giacobbe, L. In situ diagnostic analysis of the XVIII century Madonna della Lettera panel painting (Messina, Italy). *Spectrochim. Acta Part A Mol. Biomol. Spectrosc.* **2020**, *228*, 117822. [CrossRef] [PubMed]
4. Fermo, P.; Mearini, A.; Bonomi, R.; Arrighetti, E.; Comite, V. An integrated analytical approach for the characterization of repainted wooden statues dated to the fifteenth century. *Microchem. J.* **2020**, *157*, 105072. [CrossRef]
5. Castro, K.; Benito, Á.; Martínez-Arkarazo, I.; Etxebarria, N.; Madariaga, J.M. Scientific examination of classic Spanish stamps with colour error, a non-invasive micro-Raman and micro-XRF approach: The King Alfonso XIII (1889–1901 "Pelón") 15 cents definitive issue. *J. Cult. Herit.* **2008**, *9*, 189–195. [CrossRef]
6. La Russa, M.F.; Ruffolo, S.A.; Belfiore, C.M.; Comite, V.; Casoli, A.; Berzioli, M.; Nava, G. A scientific approach to the characterisation of the painting technique of an author: The case of Raffaele Rinaldi. *Appl. Phys. A* **2014**, *114*, 733–740. [CrossRef]
7. Edwards, H.G.M.; Jorge Villar, S.E.; Eremin, K.A. Raman spectroscopic analysis of pigments from dynastic Egyptian funerary artefacts. *J. Raman Spectrosc.* **2004**, *35*, 786–795. [CrossRef]
8. Vizárová, K.; Reháková, M.; Kirschnerová, S.; Peller, A.; Šimon, P.; Mikulášik, R. Stability studies of materials applied in the restoration of a baroque oil painting. *J. Cult. Herit.* **2011**, *12*, 190–195. [CrossRef]
9. Andreotti, A.; Izzo, F.C.; Bonaduce, I. Archaeometric Study of the Mural Paintings by Saturnino Gatti and Workshop in the Church of San Panfilo—Tornimparte (AQ). The Study of Organic Materials. *Appl. Sci.* **2023**. *submitted*.
10. Spoto, S.E.; Paladini, G.; Caridi, F.; Crupi, V.; D'Amico, S.; Majolino, D.; Venuti, V. Multi-Technique Diagnostic Analysis of Plasters and Mortars from the Church of the Annunciation (Tortorici, Sicily). *Materials* **2022**, *15*, 958. [CrossRef]
11. Bonizzoni, L.; Caglio, S.; Galli, A.; Lanteri, L.; Pelosi, C. Materials and Technique: The First Look at Saturnino Gatti. *Appl. Sci.* **2023**. *submitted*.
12. Bonizzoni, L.; Caglio, S.; Galli, A.; Germinario, C.; Izzo, F.; Magrini, D. Identifying Original and Restoration Materials through Spectroscopic Analyses on Saturnino Gatti Mural Paintings: How Far a Non-Invasive Approach can Go. *Appl. Sci.* **2023**. *submitted*.
13. Armetta, F.; Giuffrida, D.; Ponterio, R.C.; Falcon Martinez, M.F.; Briani, F.; Pecchioni, E.; Santo, A.P.; Ciaramitaro, V.C.; Saladino, M.L. Looking for the Original Materials and Evidence of Restoration at the Vault of the San Panfilo Church in Tornimparte (AQ). *Appl. Sci.* **2023**. *submitted*.
14. Ricci, S. Tornimparte, a Mimesis of Florence in Abruzzo. New Insights into Saturnino Gatti's Art. *Appl. Sci.* **2023**. *submitted*.
15. Germinario, L.; Giannossa, L.C.; Lezzerini, M.; Mangone, A.; Mazzoli, C.; Pagnotta, S.; Spampinato, M.; Zoleo, A.; Eramo, G. Petrographic and Chemical Characterization of the Frescoes by Saturnino Gatti (Central Italy, 15th Century): Microstratigraphic Analyses on Thin Sections. *Appl. Sci.* **2023**. *submitted*.
16. Comite, V.; Bergomi, A.; Lombardi, C.A.; Fermo, P. Characterization of Soluble Salts on the Frescoes by Saturnino Gatti in the Church of San Panfilo in Villagrande di Tornimparte (L'Aquila). *Appl. Sci.* **2023**. *submitted*.
17. Galli, A.; Alberghina, M.F.; Re, A.; Magrini, D.; Grifa, C.; Ponterio, R.C.; La Russa, M.F. Special Issue: Results of the II National Research Project of AIAr: Archaeometric Study of the Frescoes by Saturnino Gatti and Workshop at the Church of San Panfilo in Tornimparte (AQ, Italy). *Appl. Sci.* **2023**. *submitted*.
18. Pouchou, J.-L.; Pichoir, F. *Quantitative Analysis of Homogeneous or Stratified Microvolumes Applying the Model "PAP."* In Electron Probe Quantitation; Springer US: Boston, MA, USA, 1991; pp. 31–75.
19. Buzgar, N.; Apopei, A.I.; Buzatu, A. Romanian Database of Raman Spectroscopy. Available online: http://rdrs.uaic.ro (accessed on 9 March 2023).
20. Lafuente, B.; Downs, R.T.; Yang, H.; Stone, N. The power of databases: The RRUFF project. In *Highlights in Mineralogical Crystallography*; Armbruster, T., Danisi, R.M., Eds.; W. De Gruyter: Berlin, Germany, 2015; pp. 1–30. ISBN 9783110417104.
21. Caggiani, M.C.; Cosentino, A.; Mangone, A. Pigments Checker version 3.0, a handy set for conservation scientists: A free online Raman spectra database. *Microchem. J.* **2016**, *129*, 123–132. [CrossRef]
22. Pigments Checker—Modern & Contemporary Art. Available online: http://chsopensource.org/tools-2/pigments-checker/ (accessed on 9 March 2023).
23. De Benedetto, G.E.; Laviano, R.; Sabbatini, L.; Zambonin, P.G. Infrared spectroscopy in the mineralogical characterization of ancient pottery. *J. Cult. Herit.* **2002**, *3*, 177–186. [CrossRef]
24. Sadtler Database for FT-IR. Available online: http://www.ir-spectra.com/sadtler/sadtler.htm (accessed on 7 March 2023).

25. Giuntini, L.; Castelli, L.; Massi, M.; Fedi, M.; Czelusniak, C.; Gelli, N.; Liccioli, L.; Giambi, F.; Ruberto, C.; Mazzinghi, A.; et al. Detectors and Cultural Heritage: The INFN-CHNet Experience. *Appl. Sci.* **2021**, *11*, 3462. [CrossRef]
26. Wegener, M.R.; Mathew, K.J.; Hasozbek, A. The direct total evaporation (DTE) method for TIMS analysis. *J. Radioanal. Nucl. Chem.* **2013**, *296*, 441–445. [CrossRef]
27. Chukanov, N.V.; Vigasina, M.F.; Zubkova, N.V.; Pekov, I.V.; Schäfer, C.; Kasatkin, A.V.; Yapaskurt, V.O.; Pushcharovsky, D.Y. Extra-Framework Content in Sodalite-Group Minerals: Complexity and New Aspects of Its Study Using Infrared and Raman Spectroscopy. *Minerals* **2020**, *10*, 363. [CrossRef]
28. Caggiani, M.C.; Acquafredda, P.; Colomban, P.; Mangone, A. The source of blue colour of archaeological glass and glazes: The Raman spectroscopy/SEM-EDS answers. *J. Raman Spectrosc.* **2014**, *45*, 1251–1259. [CrossRef]
29. Crupi, V.; Fazio, B.; Fiocco, G.; Galli, G.; La Russa, M.F.; Licchelli, M.; Majolino, D.; Malagodi, M.; Ricca, M.; Ruffolo, S.A.; et al. Multi-analytical study of Roman frescoes from Villa dei Quintili (Rome, Italy). *J. Archaeol. Sci. Rep.* **2018**, *21*, 422–432. [CrossRef]

Disclaimer/Publisher's Note: The statements, opinions and data contained in all publications are solely those of the individual author(s) and contributor(s) and not of MDPI and/or the editor(s). MDPI and/or the editor(s) disclaim responsibility for any injury to people or property resulting from any ideas, methods, instructions or products referred to in the content.

Article

Petrographic and Chemical Characterization of the Frescoes by Saturnino Gatti (Central Italy, 15th Century)

Luigi Germinario [1], Lorena C. Giannossa [2,3], Marco Lezzerini [4], Annarosa Mangone [2,3], Claudio Mazzoli [1], Stefano Pagnotta [4], Marcello Spampinato [4], Alfonso Zoleo [5] and Giacomo Eramo [3,6,*]

1. Department of Geosciences, University of Padua, 35131 Padova, Italy; luigi.germinario@gmail.com (L.G.); claudio.mazzoli@unipd.it (C.M.)
2. Department of Chemistry, University of Bari Aldo Moro, 70125 Bari, Italy; lorenacarla.giannossa@uniba.it (L.C.G.); annarosa.mangone@uniba.it (A.M.)
3. Interdepartmental Center Research Laboratory for the Diagnostics of Cultural Heritage, University of Bari Aldo Moro, 70125 Bari, Italy
4. Department of Earth Sciences, University of Pisa, 56126 Pisa, Italy; marco.lezzerini@unipi.it (M.L.); stefano.pagnotta@unipi.it (S.P.); marcello.spampinato@unipi.it (M.S.)
5. Department of Chemical Sciences, University of Padua, 35131 Padova, Italy; alfonso.zoleo@unipd.it
6. Department of Geoenvironmental and Earth Sciences, University of Bari Aldo Moro, 70125 Bari, Italy
* Correspondence: giacomo.eramo@uniba.it

Abstract: This study presents the petrographic and chemical characterization of the frescoes in the Church of San Panfilo in Tornimparte (AQ, Italy) by Saturnino Gatti, a prominent painter of the late 15th–early 16th century, known for his exquisite technique, composition, and use of color. The characterization of the frescoes is essential for understanding the materials and techniques used by Gatti, as well as for identifying the stratigraphy and painting phases. Eighteen samples were collected from the original paint layers, later additions (17th century), and restored surfaces, and analyzed by optical microscopy, cathodoluminescence microscopy, scanning electron microscopy (SEM-EDS), µ-Raman, and electron paramagnetic resonance (EPR). The analyses revealed a microstratigraphy often made of three main layers: (1) preparation, consisting of lime plaster and sand; (2) pigmented lime, applied by the fresco technique; and (3) additional pigmented layer on the surface. The most often recurring pigments are black, red, yellow (all generally linked with the fresco technique), and blue (applied "a secco"). The presence of two painting phases was also noted in one sample, probably resulting from a rethinking or restoration. These findings contribute to the understanding of the history and past restoration works of this cultural heritage site, providing important insights not only for conservators and restorers, but also for a broader understanding of Italian fresco painting and art history of the late 15th and early 16th centuries.

Keywords: lime plaster; fresco painting; pigments; paint stratigraphy; mineralogy; petrography; SEM-EDS; Raman; EPR; Tornimparte

1. Introduction

The Italian painter Saturnino Gatti was known for his exceptional technique, composition, and use of color in his frescoes during the late 15th and early 16th centuries [1]. The Church of San Panfilo in Villagrande di Tornimparte (AQ, central Italy) is home to a significant example of his work, which is of considerable importance in the context of Italian cultural heritage [2]. The church is located in an area hit by the catastrophic earthquake of L'Aquila in 2009, but did not suffer any serious damage, contrary to many other buildings of historical and cultural significance. Studying the building, along with the artworks contained in it, will aid in its conservation and restoration [3,4].

Thanks to modern technologies, the characterization and reconstruction of altered or damaged painted surfaces can be achieved in various ways, ranging from techniques for the restitution of morphology and color [5,6] to mineralogical–petrographic and chemical characterization techniques [7–9], frequently with non-destructive [10,11] and non-invasive [12–18] methodologies.

The study of the pictorial cycle in the Church of San Panfilo was conducted through a petrographic, mineralogical, and chemical approach. The investigation aimed to identify the stratigraphy, microstructure, and composition of the plasters [19], paint layers [20], and painting techniques [7,21,22]. The study also explored the presence of compounds possibly related to the alteration processes of the frescoes and their components [23]. The analyses may provide useful indications for a comprehensive understanding of the artistic technique of Saturnino Gatti and the conservation of the frescoes. The upcoming restoration of the church will be based on the principles of reversibility, compatibility, and durability. In this context, a multi-analytical approach, as demonstrated by Bersani et al. [24], Romani et al. [25], and Alberghina et al. [26], is extremely effective in providing information, from the macro to the micro domain [27], not only on pigments but also on specific binders (such as egg whites, oils, and resins) and alteration patinas [28]. The combination of this information can guide the restorer in the decision and implementation of specifically targeted conservation practices, useful for restoring the condition and readability of the artwork in case of ordinary and extraordinary maintenance [29–31], even with bioremediation techniques [32].

This paper is part of the Special Issue "Results of the II National Research Project of AIAr: archaeometric study of the frescoes by Saturnino Gatti and workshop at the Church of San Panfilo in Tornimparte (AQ, Italy)" [33], which presents the scientific results of the II National Research Project conducted by members of the Italian Association of Archaeometry (AIAr). The aim of the project was to conduct an in-depth study of the frescoes by Saturnino Gatti and his workshop, with the objective of contributing to a broader understanding of Italian fresco painting and art history of the late 15th and early 16th centuries. For a more detailed overview of the project's objectives, the readers can refer to the introduction of the Special Issue [33].

2. Materials and Methods

2.1. Sampling

The sampling campaign was carried out on 1 March 2021, under the supervision of the officials of Soprintendenza Archeologia, Belle Arti e Paesaggio for the provinces of L'Aquila and Teramo. A total of 18 out of 47 samples collected from the paint layers and plasters of the frescoes were analyzed by this research group. Table 1 shows their original location and brief description. Some consist of multiple specimens; in that case, the alphanumerical ID is associated with a letter with alphabetical progression. The sampling areas considered here are the following (Figure 1): Panel A (The Kiss of Judas and the Capture of Christ); Panel B (Flagellation and Crowning with Thorns/Christ at the Column—a depiction that was extensively reworked, probably during the 17th century); Panel C (central wall of the apse with a window opened in 1926; originally, the Crucifixion was depicted there, a fragment of which still remains); Panel D (Lamentation over the Dead Christ); Panel E (Resurrection); and Vault (God the Father in Glory among angels and saints).

Figure 1. Sampling spots with associated sample ID.

Table 1. Source and macroscopic features of the analyzed samples (the question marks indicate doubtful interpretations requiring in-depth validation). The analytical techniques applied for each sample ("t.s." = thin section; "f." = fragment with no prior preparation) are also specified (details follow in Section 2.2).

Sample	Panel	Description	Investigation Technique
SG9A	A	Plaster with pale ochre paint	OM, SEM-EDS (t.s.)
SG9B	A	Plaster with pale ochre paint and decay products (?)	OM, SEM-EDS (t.s.)
SG10	A	Plaster with violaceous paint	OM, SEM-EDS (t.s.)
SG11	B	Plaster with dark-brown paint	OM, SEM-EDS (t.s.)
SG12	B	Plaster with blue–gray paint and makeover (?)	OM, SEM-EDS (t.s.)
SG13	C	Plaster with white paint	OM, SEM-EDS (t.s.)
SG14	E	Plaster with dark paint and protective coating (?)	OM, SEM-EDS (t.s.)
SG15	E	Plaster with blue layer	OM, SEM-EDS (t.s.)
SG16	A	Pale-green paint and plaster (reworked area?)	OM, CL, SEM-EDS, µ-Raman (t.s.)
SG17	A	Porphyry-like paint and plaster	OM, CL, SEM-EDS, µ-Raman (t.s.)
SG23A	A	Plaster	OM, CL, SEM-EDS, µ-Raman (t.s.)
SG25	D	Pale-blue paint on a red–brown layer (morellone?) and plaster	OM, CL, SEM-EDS, µ-Raman (t.s.)
SG35	D/E	Green layer	EPR (f.)
SG36	D/E	Light-blue layer	EPR (f.)
SG39B	Vault	Blue paint on a red–brown layer (morellone?) and plaster	OM, CL, SEM-EDS, µ-Raman (t.s.)
SG40	Vault	Filling mortar	OM, CL, SEM-EDS, µ-Raman (t.s.)
SG46	C	Plaster with remains of a brown preparation layer	OM, CL, SEM-EDS, µ-Raman (t.s.)
SG47	A	Plaster with remains of a brown preparation layer	OM, CL, SEM-EDS, µ-Raman (t.s.)

2.2. Methods

In Table 1, the analytical methods applied to the samples are reported for identifying the stratigraphy and petrographic characteristics of the plaster, the composition of the binder, the pigments, and the alteration products.

Cross-sections for the microstratigraphic analysis were prepared from sixteen samples (SG9A, SG9B, SG10, SG11, SG12, SG13, SG14, SG15, SG16, SG17, SG23A, SG25, SG39B, SG40, SG46, SG47) at the Department of Earth Sciences at the University of Pisa. The observation of the thin sections was performed using a ZEISS AxioPlan polarized light microscope (OM). For the preparation of the cross-sections, the samples were first observed under a ZEISS STEMI 305 stereo microscope in order to decide the position and direction of the cut and highlight all the layers of the samples, from the surface to the deepest part.

A first set of eight samples (SG9A-B, SG10, SG11, SG12, SG13, SG14, SG15) was investigated at the Department of Geo–environmental and Earth Sciences of the University of Bari Aldo Moro, with the following instrumentation and analytical conditions.

Scanning electron microscope (SEM) observations were made on thin sections, previously fixed on aluminum specimen holders and coated with graphite. An SEM EVO-50XVP (LEO), equipped with an Oxford Instruments AZTEC EDS microanalysis system with SD X-Max^N detector (80 mm^2), was used. The accuracy of the analytical data was verified using various standards produced by Micro-Analysis Consultants Ltd. (St. Ives, UK). The working voltage was 15 kV and the beam current between 250 and 400 pA. Spectra acquisitions lasted 50 s, with counts ranging from 25,000 to 30,000. Chemical maps were acquired with a dwell time of 100 µs, counting time of 10 min, and a resolution of 2048.

A second set of ten samples was analyzed at the University of Padova, eight at the Department of Geosciences (SG16, SG17, SG23A, SG25, SG39B, SG40, SG46, and SG47), and two, without any preliminary preparation, at the Department of Chemical Sciences (SG35 and SG36).

The samples investigated at the Department of Geosciences were prepared as polished stratigraphic thin sections and analyzed by cathodoluminescence microscopy (CL), µ-Raman, and SEM after C coating. The CL analyses were performed with an optical microscope Nikon Labphot2 Pol (with long-working-distance objectives providing 4×, 10×, and 40× magnification) equipped with a CCL (Cold Cathode Luminescence) 8200 mk3; in standard analytical conditions, voltage is 20 kV and beam current is about 200 µA. For the µ-Raman analyses, the microscope used was a Raman Thermo Scientific DXR with 532 nm laser, 50× long-working-distance objective, 3 mW power, 25 µm pinhole, and frequency range between 100 and 3500 cm^{-1}. The SEM analyses were performed with a microscope CamScan MX2500 with W source, detector of secondary and back-scattered electrons, and EDAX EDS (energy-dispersive spectroscopy) microanalysis; the analytical conditions were 25 kV voltage and 15 to 25 mm working distance.

The samples investigated at the Department of Chemical Sciences were analyzed by X-Band CW-EPR spectroscopy. The fragments were placed in a quartz tube as sample holder (3 mm inner diameter, 4 mm outer diameter) and examined at room temperature with a Bruker ECS106 instrument, equipped with a TE102 cavity. Spectra were acquired with 10 scan, modulation amplitude of 0.5 mT, field sweep of 160 mT, and center field of 320 mT. Microwave power was set to 20 mW and microwave frequency was 9.538 GHz. A short µ-Raman characterization of the back and front of the two samples was carried out using a Renishaw InVia Spectrometer with laser at 514 nm, objectives 10× or 20×, 10 s integration time, 50% total output power.

3. Results

3.1. Petrography (OM, CL)

The photomicrographs of all the thin sections analyzed are included in Figures 2 and 3. The preparation layer of the painted surface appears to be made up, in all the samples analyzed, of a lime and sand plaster, on which is superimposed a layer of lime, probably applied "a fresco", pigmented with yellow and red ochres. In the case of red ochres, these are always associated with the presence of C black, while this association is sporadic in yellow ochres. The pigmented layer has a thickness between 7 and 30 µm in all samples except for sample SG11, where it exceeds 75 µm. From the analyzed cross-sections, the primary color application is carried out with red and yellow ochre also containing a C black pigment. This draft was given with the fresco technique. Subsequently, other colored layers were added (green, yellow, blue, and white), in which the use of a "tempera" technique is likely.

Sample SG40 bears witness to a phase of remaking/rethinking where the previous draft with a very thin layer of red ochre and C black is at the base, which is superimposed on a thin layer pigmented in blue and is covered with a plaster free of aggregate on which there is a new color draft with yellow ochres. Inside the *intonachino* that separates the two phases, there are some granules of a blue pigment and red ochre.

The CL analyses supported the previous observations by providing elements for characterizing the plaster substrates and discriminating the composition of binder and aggregate based on the characteristic luminescence of the different minerals. The observations pointed out that the plaster is generally composed of a lime binder with aggregates of quartz, alkali–feldspars, and plagioclases in various proportions, with minor amounts of carbonate minerals. The aggregate is, in most cases, moderately to poorly sorted.

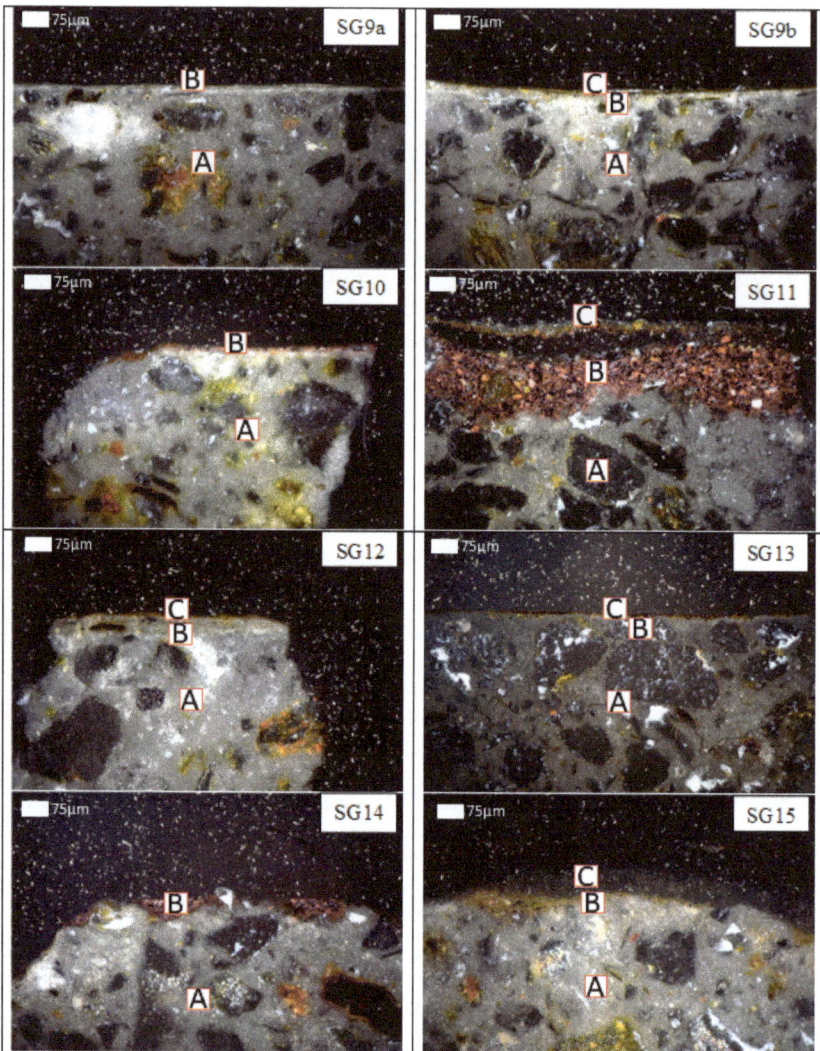

Figure 2. Microphotographs of thin sections of the samples SG9a, SG9b, SG10, SG11, SG12, SG13, SG14, and SG15. The capital letters "**A**", "**B**", and "**C**" indicate the different layers visible in cross-section, with "**A**" being the innermost (the plaster) and the others (mostly pigmented) ordered by decreasing depth.

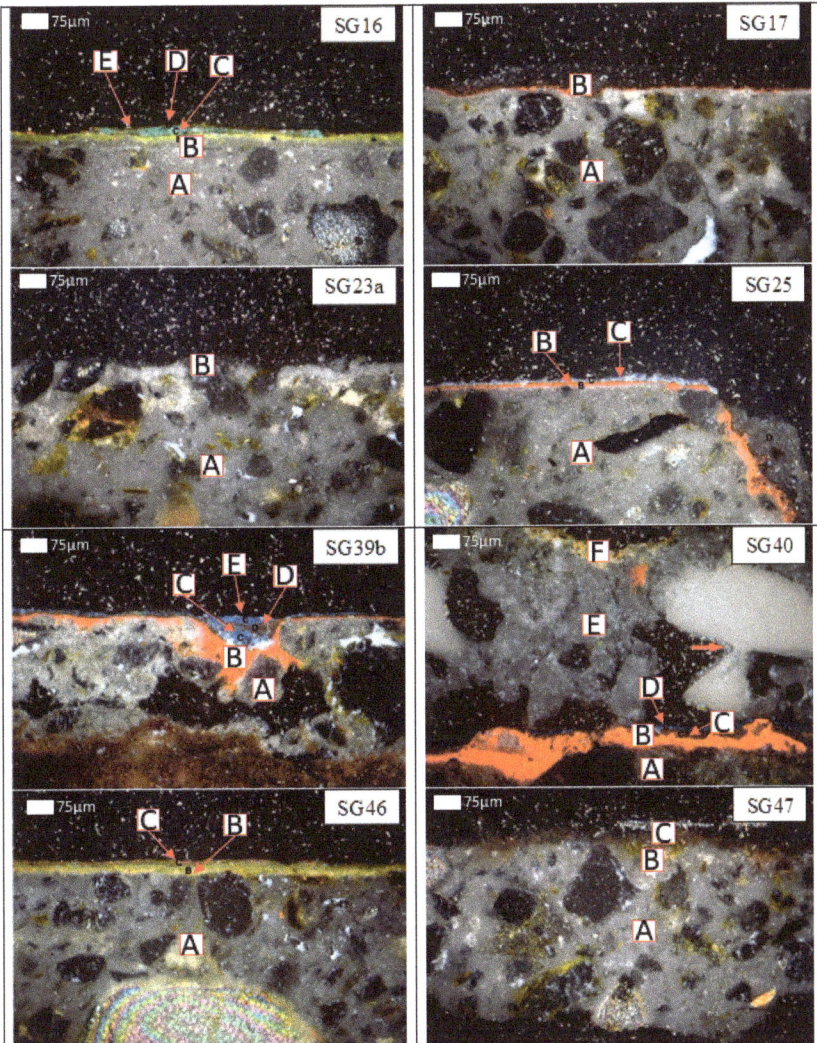

Figure 3. Microphotographs of thin sections of the samples SG16, SG17, SG23a, SG25, SG39b, SG40, SG46, and SG47. The capital letters "**A**", "**B**", "**C**", "**D**", "**E**", and "**F**" indicate the different layers visible in cross-section, with "**A**" being the innermost (the plaster) and the others (mostly pigmented) ordered by decreasing depth. Sample SG40 shows a break between the layers consisting of highly purified plaster (**E**) in which remnants of a blue pigment can be seen (indicated by the red arrow), probably due to the use of a dirty brush or the mixing of pigments from the underlying layer (**D**).

3.2. SEM-EDS and µ-Raman

The paint layers of the three samples from Panel A (SG16, SG17, and SG47) share the same main pigment, that is, ochre, either red (hematite) and/or yellow (goethite), applied on the plaster "a fresco". This is suggested by the absence of microstructural discontinuities between the plaster and the paint traces. Generally, the ochres are mixed with fine particles of C black. The stratigraphy of sample SG16 is worthy of further remarks since it is the only one showing "a secco" layers, with additional pigments applied on the dry plaster (Figure 4): one green–blue middle layer made of malachite and one yellow–brown surface layer probably composed of Cr yellow (because of the detection of Pb chromate grains); this last identification, however, needs to be taken with reservation. Finally, the SEM-EDS analyses of sample SG47 also highlighted the presence of gypsum particles and several surface irregularities and detachments.

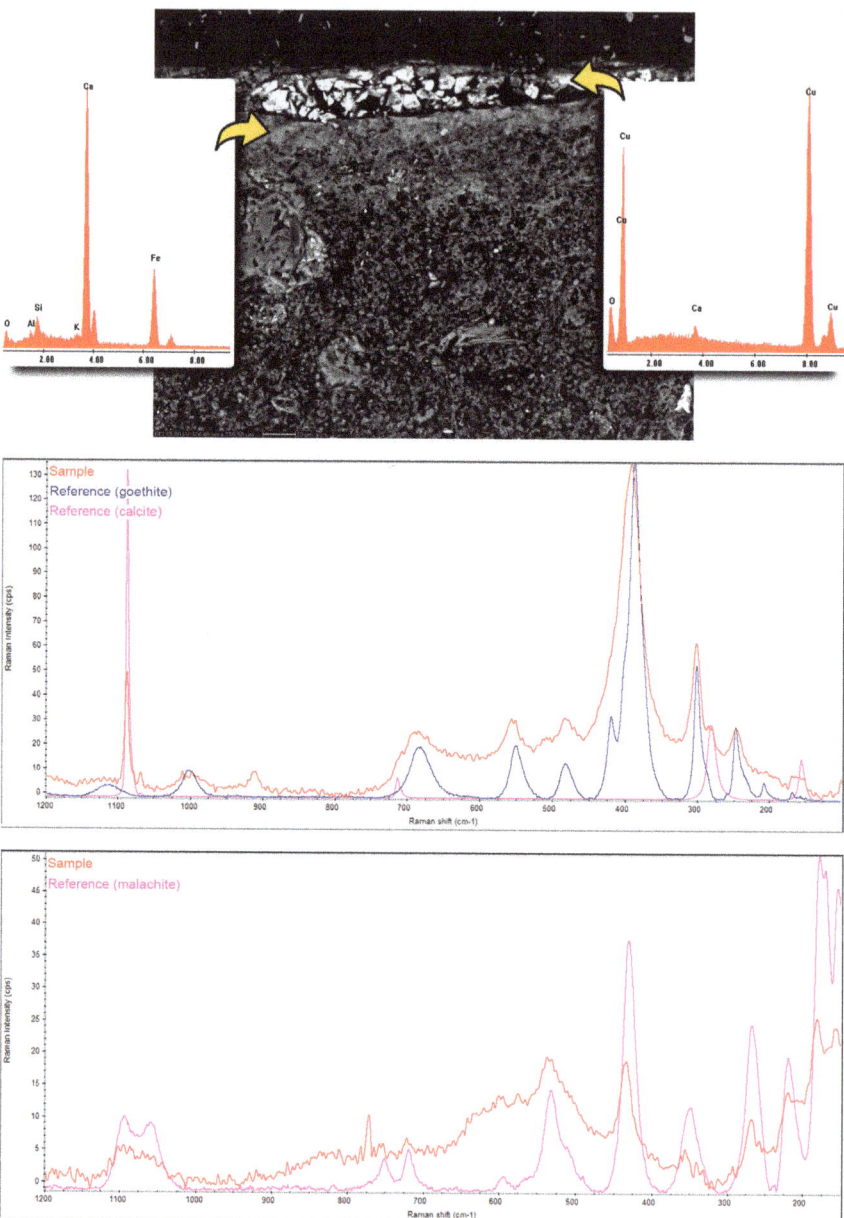

Figure 4. Sample SG16: SEM-BSE image with associated SEM-EDS spectra from the spots indicated, and μ−Raman spectra calculated on two different paint layers (actual analyses in red, reference spectra from mineral databases in blue and pink).

The SEM-EDS analysis revealed similar microstructural and compositional characteristics for samples SG9A and SG9B. The samples are layered, with an air lime-based arenaceous plaster (A) on which two paint layers (B, ca. 20 μm; C, ca. 8 μm) are visible (Figure 5). A detachment is observed between layers B and C. Layer B is more compact than C and both are air lime-based. The EDS microanalysis of the small areas of the painted layers reports the presence of iron oxyhydroxides in layer B, and calcium phosphate (bone black), zinc oxide, and iron oxyhydroxides as pigments in layer C. The compositional traverse in Figure 5 shows the gradual diffusion of phosphorus in the underlying layer, confirming the hypothesis of an "a fresco" painting. The correlation between K, Al, Si, and Fe in layer C might indicate the presence of red bolus [34]. S is due to the presence of gypsum.

Additionally, in SG10, two lime-based layers are visible. Layer B is ca. 20 μm thick and pigmented with hematite. On layer B, an additional pigmented layer (C) partially detached from the previous was recognized, with a thickness of less than 10 μm (Figure 6).

The two samples of Panel B (SG11 and SG12) show two painted layers applied on an air lime-based arenaceous plaster (A). In SG11, layer B can be distinguished by its content of calcium phosphate, zinc oxide, red ochre, and gypsum. Replicated measurements on two other detail areas on layers B and C of the same fragment provided similar data. The binder in layer C is probably of organic origin.

Sample SG12 is layered, with an air lime-based arenaceous plaster (A) on which a B layer is visible in adhesion and a laterally discontinuous C layer. The presence of Mn in the Fe-rich particles in layer B may indicate the use of Sienna, whereas layer C is a lime paint with red ochre. Some gypsum was also identified in this last layer.

In Panel C, SG13 is stratified, with an arenaceous lime plaster (A) on which there are two painted layers. Layer B has a gradual stratigraphic contact with A and sharp contact with C. The EDS chemical mapping shows a high P content in layer C due to some calcium phosphate. Ochre is present in both layer B and layer C.

The painted surface of sample SG46 is also composed of ochre; in this case, however, the yellow ochre (goethite) and red ochre (hematite) are not mixed, with the former composing the "a fresco" bottom layer, and the latter, the upper "a secco" layer, also including scattered C black particles. The upper layer also recorded SEM signals of S and Cl. From a microstructural point of view, the sample shows a distinct surface microcracking, with fractures expanding across the paint layers or partially separating them from the plaster substrate.

The sample from Panel D (SG25) again has red ochre (hematite with some parts of magnetite) as the pigment originally used for the "a fresco" layer. The upper layer, probably applied "a secco", is blue and made of a combination of ultramarine and Ti white, suggested by the detection of the phases lazurite and rutile; some broad Raman peaks (from about 1300 to 1500 cm^{-1}) also point out the presence of organic matter. The SEM microanalyses also revealed the presence of baryte. Finally, another layer, differently oriented in respect to the others (that is, not observed on the surface but sinking into the plaster), is formed of S and Ca (gypsum?) and covers an inner thin red film; although this has the same color as the red ochre layer and is also in direct contact with the substrate, its composition is slightly different, with frequent SEM-EDS signals of Fe, Pb, and Ba from baryte.

SG14 and SG15 were sampled from Panel E. In the first case, layer B is present discontinuously on layer A. The stratigraphic contact is sharp, with complex geometry. The binder of the plaster (A) is air lime. Chemical mapping reveals the presence of Ca phosphate, Zn oxide, and ochre (Figure 7). In the second case, layer B identified by optical microscopy shows no clear separation from layer A. There, Fe enrichment and carbonaceous inclusions are detected. The outer surface of B also shows the presence of gypsum. Layer C is poorly preserved and consists of gypsum and calcite.

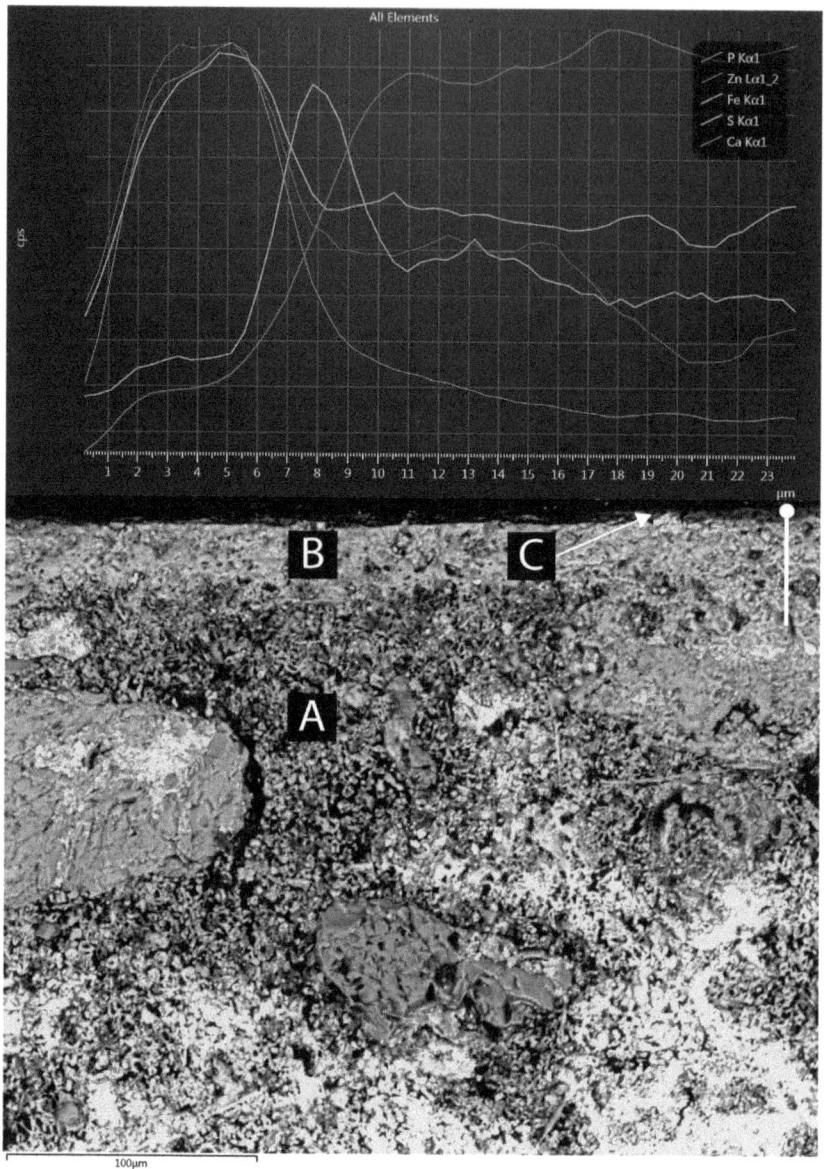

Figure 5. SG9A, SEM-BSE image (**bottom**) of the outer portion of the sample (plaster layer "**A**" and paint layers "**B**" and "**C**") and the EDS traverse across the identified layers (**top**). The intensities of the K and L lines of the elements in the legend are normalized to better detect lateral variation in the painted layers.

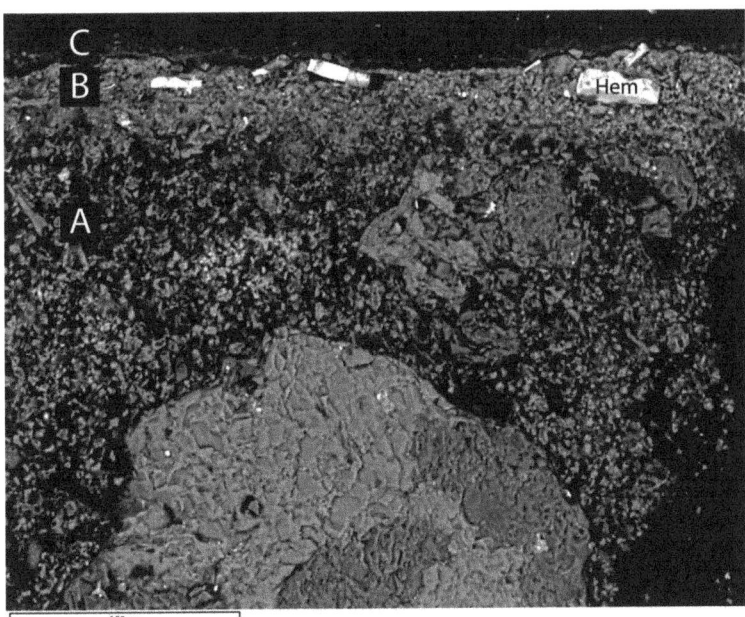

Figure 6. SG10, SEM-BSE image of the outer portion of the sample (plaster layer "**A**" and paint layers "**B**" and "**C**"). Light grey tabular fragments of hematite (Hem) occur as pigment in layer **B**.

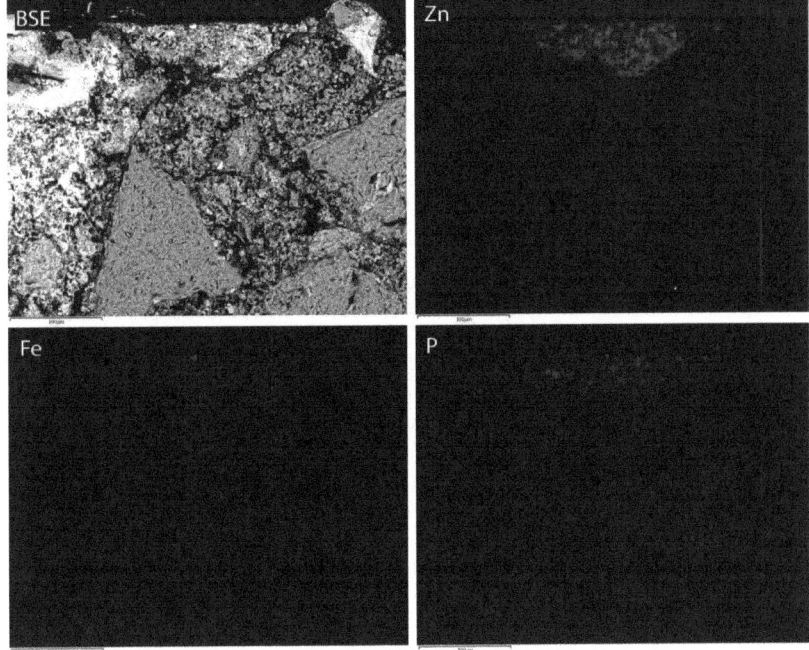

Figure 7. SG14, SEM-BSE image of the outer portion of the sample and related EDS maps of Zn, Fe, and P.

The sample from Panel "sky" (SG23A) is a non-painted plaster fragment.

The samples from the vault (SG39B and SG40) are the most complex compositionally, considering their stratigraphy composed of six layers. Sample SG39B has four different paint layers above the plaster substrate: the first and deepest, applied "a fresco", is made of red ochre (hematite with magnetite); the second, applied "a secco", is blue and made of azurite (with traces of Zn and As)—it is detectable only in a very small area and never on the exposed surface; the third, more continuous but almost colorless, has an ambiguous composition, enriched in gypsum (from the SEM signals of S and Ca) and resins or organic polymers (from the μ-Raman spectra); finally, the fourth, the most superficial, covers the entire surface and is blue and composed of ultramarine (lazurite), with traces of Pb particles (pointing out the possible mix of ultramarine with Pb white) and Ba (probably baryte) (Figure 8). With regard to sample SG40, this is formed by two main stratigraphic domains. The first has a sequence of three layers overlying the plaster substrate: from bottom to top, one red "a fresco" film composed of cinnabar, one layer rich in gypsum and resins or organic polymers, and one blue "a secco" layer with ultramarine (lazurite) mixed with Pb white (basic Pb carbonate) with baryte traces. The second overlaying stratigraphic domain consists of an additional lime plaster layer (much more porous and with nearly just lime, without the silicate aggregate) and one final yellow–brown paint film made of yellow ochre (goethite) mixed with C black grains, applied "a fresco" on the surface. The surface also records the presence of gypsum.

3.3. EPR

CW-EPR was applied on two small fragments (SG35 and SG36) to evaluate if this technique could provide further information on specific aspects of the samples; EPR is able to spot paramagnetic defects (color centers, radicals) or paramagnetic metals (Cu^{2+}, Mn^{2+}, Fe^{3+}), which are commonly present in many pigments and materials as substitutional ions. EPR is a bulk technique, so the whole sample is examined and specific pieces of information related to the different painted layers are not accessible.

A preliminary overview of the two fragments, front and back, was conducted using optical microscopy and Raman spectroscopy.

The SG35 front (Figure 9) is characterized by a layer with small blue grains on a yellow background giving a green hue. Some bigger blue grains can be spotted, which Raman identifies unambiguously as azurite. On the back, pale green crystals are present, which Raman identifies likely as a combination of brochantite (basic copper sulfate) and atacamite (basic copper chloride), which are likely alteration by-products from an original copper-based pigment. Two unidentified peaks at 444 and 1475 cm^{-1} suggest the presence of some other compound.

The EPR spectrum of SG35 (Figure 10) is characterized by a sharp signal, whose g-factor (g = 2.0035) is related to a C radical, likely a radical in C black. A six-line pattern is also observed, attributable to Mn^{2+} in Ca carbonate, and a signal at g~2.06, with linewidth Hpp = 4.3 mT can be attributed to Cu^{2+} in some compounds deriving from the alteration of a copper-based pigment. In azurite, brochantite, and atacamite (the copper-based pigments detected by Raman), the Cu^{2+} ions are either antiferromagnetically coupled or have very broad EPR lines; therefore, the g~2.06 signal is unlikely to be related to these latter and could be attributed to some complex of copper with an organic compound.

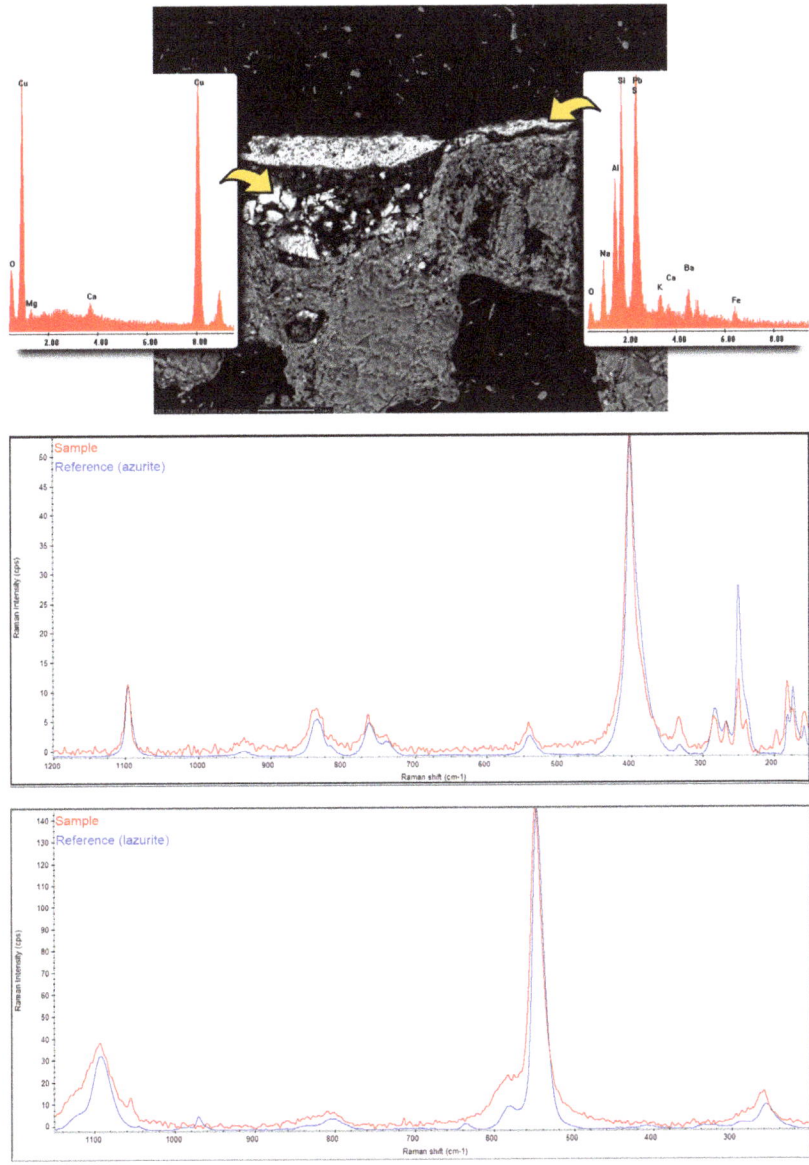

Figure 8. Sample SG39B: SEM-BSE image with associated SEM-EDS spectra from the spots indicated, and μ-Raman spectra calculated on two different paint layers (actual analyses in red, reference spectra from mineral databases in blue).

Figure 9. **Above**: SG35 front (**left**) and back (**right**). **Below**: SG36 front (**left**) and back (**right**).

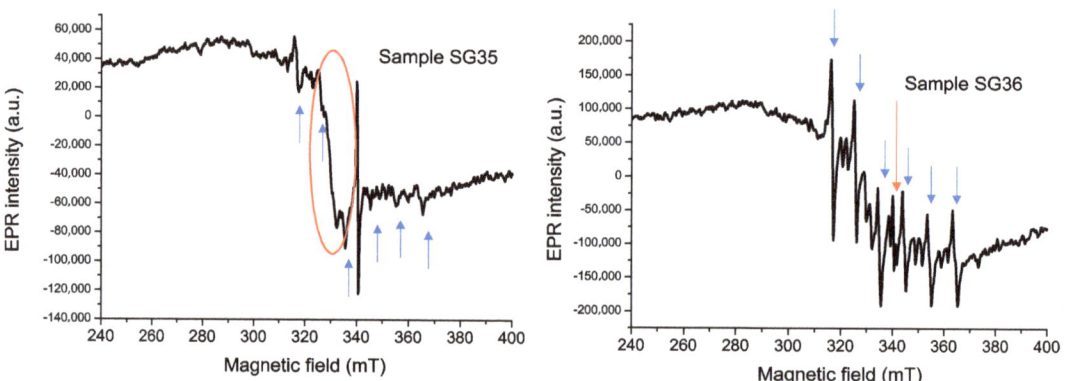

Figure 10. Left: EPR of SG35. Red asterisk marks the sharp signal of a C radical; blue arrows mark the sextet of Mn^{2+}; the red circle highlights the signal of Cu^{2+}. **Right**: EPR of SG36. Red arrow marks a C radical; blue arrows mark the sextet of Mn^{2+}.

The SG36 front is uniformly covered by blue grains, which Raman identifies again as azurite. The SG36 back is yellow–red on a white background; only peaks due to calcium carbonate are observed in the Raman spectrum.

In the EPR spectrum of sample SG36, no Cu^{2+} signal is observed, while a strong Mn^{2+} six-line pattern is seen, again attributable to manganese in Ca carbonate. The signal is stronger than in SG35. The C black radical signal is also seen, although weaker than in SG35. Both SG35 and SG36 present a broad background signal attributable to iron oxyhydroxides.

4. Discussion

4.1. Stratigraphy: Original and Successive Interventions

The samples analyzed showed stratigraphic and compositional similarities, summarized in Table 2. A common substratum of lime plaster with similar composition was identified in all the samples. The paint layers show a more or less complex stratification according to the pictorial intention of the artist and the further restorations.

Table 2. Summary of the results obtained. The question marks indicate doubtful interpretations requiring in-depth validation. Pigments of past restorations are reported in italics.

Sample	Layer A (Plaster)	Layer B	Layer C	Layer D	Layer E	Layer F
SG9A		red bolus	bone black	/	/	/
SG9B		red bolus	*bone black, red bolus, ZnO*	/	/	/
SG10		hematite, C black	*red bolus, ZnO*	/	/	/
SG11		bone black, red ochre, ZnO	organic?	/	/	/
SG12	Air lime with quartz, alkali–feldspars, and plagioclases, micas with minor carbonates as aggregate	C black, sienna	/	/	/	/
SG13		C black, ochre	bone black, ochre	/	/	/
SG14		bone black + ochre + ZnO	/	/	/	/
SG15		C black + ochre	/	/	/	/
SG16		yellow ochre + C black	malachite	*Cr yellow?*	/	/
SG17		red ochre + C black	/	/	/	/
SG23A		/	/	/	/	/
SG25		red ochre	*ultramarine + Ti white*	/	/	/
SG35		azurite [1]	/	/	/	/
SG36		azurite	/	/	/	/
SG39B		red ochre	azurite	resins/org. polymers?	*ultramarine + Pb white*	/
SG40		cinnabar	resins/org. polymers?	*ultramarine + Pb white*	*lime plaster*	*yellow ochre + C black*
SG46		yellow ochre	red ochre + C black	/	/	/
SG47		yellow ochre + C black	red ochre	/	/	/

[1] Altered in brochantite/atacamite.

The stratigraphy shows a combination of "a fresco" and "a secco" techniques [35,36] in almost all the samples, as supported by the thickness of the layers and carbonation of the surfaces [37,38]. Further details about the technology are reported in Section 4.2. Most of the samples show traces of past restoration (Table 2), as also observed by other authors in different samples of the same panels [39–42].

Other compounds not relatable to actual pigments or dyes are the resins and organic polymers often detected in thick layers on the surface of the samples. These may trace back to finishing touches or restoration works [43].

4.2. Lime Plasters: Raw Materials and Technological Aspects

The plasters (layer A) are classifiable as air lime (CaO > 97 wt. %) mortars with an arenaceous texture. The absence of calcination relics and slacking nodules points to an accurate preparation of the lime putty. The aggregate consists of mono- and polycrystalline quartz, alkali–feldspars, plagioclases, micas, Fe aggregates, and minor carbonates. The texture is unimodal to seriate, mainly in the range of fine sand. These petrographic features point to fluvial fine sand (Torrente Raio?) used as aggregate (Figures 2 and 3) and are compatible with the siliciclastic composition of the turbiditic complex in the area of Tornimparte [44]. The lime/aggregate ratio of about 1:1 allowed workability and a smoother and denser plaster. Contrary to what is reported in various medieval treatises, the plaster intended to receive the color does not contain marble or spathic calcite powder as part of the aggregate [45].

4.3. Pigments, Binders, and Painting Techniques

Red and yellow ochres were found as the main pigments in the deeper pictorial layers in most samples (layers B, C: samples SG9-SG15, SG16, SG17, SG25, SG39B, SG46, and SG47). These layers were likely applied "a fresco", as indicated by the analysis of the SG16 sample. C black was also detected in many of these inner layers. External pictorial layers (C, D) often presented more recent pigments of different colors, sometimes applied "a secco". For example, malachite and maybe Pb chromate were detected in SG16, ultramarine and Ti white in SG25, ultramarine and azurite in SG39B, and ultramarine and Pb white in SG40. In many samples, the occurrence of Ca sulfate, likely an alteration product, was inferred by the presence of S and Ca in the SEM-EDS analysis (Section 4.3). In addition, the presence of polymeric organic compounds was supposed by the Raman analyses in two samples (SG40, SG25). It is worth noting that SG9 and SG10 have a layer containing Ca phosphate (bone black) and Zn oxide present as probable materials from later interventions. In SG35, the EPR analysis detects signals due to C black radicals and copper alteration products; in SG36, the EPR signal was detected due to Mn (II) in Ca carbonate.

The frequent presence of ochre and C black in fresco layers may be attributed to the primary design made with only red and yellow ochre (e.g., sample SG47), to which layers of other colors are superimposed (blue, green, yellow, black, and white), made with an "a secco" technique. In particular, the use of C black is typical for the dusting technique, for the realization of the sketch of the drawing, or to darken the hue of other colors.

Some microchemical characteristics observed on the stratigraphic thin sections allow for making hypotheses on Saturnino Gatti's preferred painting technique when working on his frescoes in Tornimparte. For example, the identification of frequent "a secco" layers, evident from the lack of blending between pigment grains and plaster and the presence of sharp but plain discontinuities, points out a two-step working technique, with the artist tending to apply color corrections on the dry plaster. In others (e.g., sample SG39B), pigment traces localized only in specific areas or covered by further, different paint films, may point out mistakes or reworks, or even second thoughts on the pigment choice—as for azurite, detected only in such contexts.

4.4. Alteration and Decay

The most recurring phase linked to the alteration of the painted walls is gypsum (e.g., samples SG25, SG39B, SG40, SG46, SG47, etc.). Minor traces of baryte and chlorides were also detected. Those alteration products may be frequently associated with surface microcracking and disintegration because of cyclic processes of salt weathering: the secondary phases crystallize and dissolve below and within the paint films, produce subflorescences, and exert localized pressures and mechanical stresses.

The EPR spectra acquired on two samples, SG35 and SG36, indicate that the technique could be useful in the identification of copper alteration by-products, deriving from chemical interaction between Cu^{2+} and organic binders (e.g., in the "a secco" azurite layers). Additionally, C radicals could be spotted. These products can have a role in fresco degradation, and therefore, their identification has some interest.

5. Conclusions

The petrographic and chemical analysis of a selection of samples from the Church of San Panfilo revealed some significant information about the technology and past restorations of Gatti's frescoes. Summing up, all the samples are characterized by a base layer of plaster and up to five pictorial layers. The layers containing ochres, raw sienna, C black, and cinnabar are related to the original fresco. The pigments associated with additions or past restorations are red bolus, bone black, ZnO, ochre, ultramarine, azurite, malachite, whites (Pb or Ti), and yellows (Pb and Cr?). The preparation of lime plaster was based on the calcination of local Mesozoic limestones and the use of siliciclastic fluvial sand. The absence of slaking lumps points to the use of aged lime putty as a binder.

These findings, seen in the broader picture of the work conducted by all the Italian research groups involved in this AIAr's national project, have a double value: they help to comprehend the past better, providing insights into the creation process and history of Saturnino Gatti's frescoes; and they preserve it knowingly in the future, giving precious scientific and technical diagnostic information for the restorations soon to begin.

Author Contributions: Conceptualization, G.E. and L.G.; investigation, G.E., M.L., S.P., L.G., C.M., M.S. and L.C.G.; resources, G.E., M.L. and C.M.; writing—original draft preparation, G.E., L.G., S.P. and A.Z.; writing—review and editing, G.E., A.M., C.M. and M.L.; visualization, G.E., S.P., L.G. and A.Z.; supervision, G.E. All authors have read and agreed to the published version of the manuscript.

Funding: This work was performed in the framework of the Project Tornimparte—"Archeometric investigation of the pictorial cycle of Saturnino Gatti in Tornimparte (AQ, Italy)" sponsored in 2021 by the Italian Association of Archeometry AIAR (www.associazioneaiar.com, accessed on 1 June 2023).

Data Availability Statement: All data are in the paper.

Acknowledgments: The authors thank Pasquale Acquafredda and Nicola Mongelli for their technical support in SEM analysis. This research benefited from instrumental upgrades of Potenziamento Strutturale PONa3-00369 of the University of Bari Aldo Moro titled "Laboratorio per lo Sviluppo Integrato delle Scienze e delle Tecnologie dei Materiali Avanzati e per dispositivi innovativi (SISTEMA)".

Conflicts of Interest: The authors declare no conflict of interest.

References

1. Arbace, L. I Volti dell'Anima. Saturnino Gatti. In *Vita e Opere di un Artista del Rinascimento*; Paolo De Siena Editore: Pescara, Italy, 2012; ISBN 88-96341-11-6.
2. Zezza, A. Paintings, Frescoes, and Cycles. In *A Companion to the Renaissance in Southern Italy (1350–1600)*; Di Vitiis, B., Ed.; Brill: New York, NY, USA, 2022; pp. 591–617. ISBN 978-90-04-52637-2.
3. Vittorini, A. L'Aquila 2009–2019: Back to the Future. Cultural Heritage and Post-Seismic Reconstruction Challenges. In *Invisible Reconstruction. Cross-Disciplinary Responses to Natural, Biological and Man-Made Disasters*; Patrizio Gunning, L., Rizzi, P., Eds.; Fringe; UCL Press: London, UK, 2022; pp. 11–28.
4. Quagliarini, E.; Lenci, S.; Seri, E. On the Damage of Frescoes and Stuccoes on the Lower Surface of Historical Flat Suspended Light Vaults. *J. Cult. Herit.* **2012**, *13*, 293–303. [CrossRef]
5. Lermé, N.; Hégarat-Mascle, S.L.; Zhang, B.; Aldea, E. Fast and Efficient Reconstruction of Digitized Frescoes. *Pattern Recognit. Lett.* **2020**, *138*, 417–423. [CrossRef]
6. Bruno, N.; Mikolajewska, S.; Roncella, R.; Zerbi, A. Integrated Processing of Photogrammetric and Laser Scanning Data for Frescoes Restoration. *Int. Arch. Photogramm. Remote Sens. Spat. Inf. Sci. ISPRS Arch.* **2022**, *46*, 105–112. [CrossRef]
7. Piovesan, R.; Maritan, L.; Amatucci, M.; Nodari, L.; Neguer, J. Wall Painting Pigments of Roman Empire Age from Syria Palestina Province (Israel). *Eur. J. Mineral.* **2016**, *28*, 435–448. [CrossRef]
8. Pecchioni, E.; Pallecchi, P.; Giachi, G.; Calandra, S.; Santo, A.P. The Preparatory Layers in the Etruscan Paintings of the Tomba dei Demoni Alati in the Sovana Necropolis (Southern Tuscany, Italy). *Appl. Sci.* **2022**, *12*, 3542. [CrossRef]
9. Mangone, A.; Colombi, C.; Eramo, G.; Muntoni, I.M.; Forleo, T.; Giannossa, L.C. Pigments and Techniques of Hellenistic Apulian Tomb Painting. *Molecules* **2023**, *28*, 1055. [CrossRef]
10. Holclajtner-Antunović, I.; Stojanović-Marić, M.; Bajuk-Bogdanović, D.; Žikić, R.; Uskoković-Marković, S. Multi-Analytical Study of Techniques and Palettes of Wall Paintings of the Monastery of Žiča, Serbia. *Spectrochim. Acta Part A Mol. Biomol. Spectrosc.* **2016**, *156*, 78–88. [CrossRef] [PubMed]
11. Miriello, D.; Bloise, A.; Crisci, G.M.; De Luca, R.; De Nigris, B.; Martellone, A.; Osanna, M.; Pace, R.; Pecci, A.; Ruggieri, N. Non-Destructive Multi-Analytical Approach to Study the Pigments of Wall Painting Fragments Reused in Mortars from the Archaeological Site of Pompeii (Italy). *Minerals* **2018**, *8*, 134. [CrossRef]
12. Garavelli, A.; Andriani, G.F.; Fioretti, G.; Iurilli, V.; Marsico, A.; Pinto, D. The "Sant'Angelo in Criptis" Cave Church in Santeramo in Colle (Apulia, South Italy): A Multidisciplinary Study for the Evaluation of Conservation State and Stability Assessment. *Geosciences* **2021**, *11*, 382. [CrossRef]
13. Carlomagno, G.M.; Meola, C. Comparison between Thermographic Techniques for Frescoes NDT. *NDT E Int.* **2002**, *35*, 559–565. [CrossRef]

14. Gusella, V.; Cluni, F.; Liberotti, R. Feasibility of a Thermography Nondestructive Technique for Determining the Quality of Historical Frescoed Masonries: Applications on the Templar Church of San Bevignate. *Appl. Sci.* **2021**, *11*, 281. [CrossRef]
15. Almaviva, S.; Fantoni, R.; Colao, F.; Puiu, A.; Bisconti, F.; Fiocchi Nicolai, V.; Romani, M.; Cascioli, S.; Bellagamba, S. LIF/Raman/XRF Non-Invasive Microanalysis of Frescoes from St. Alexander Catacombs in Rome. *Spectrochim. Acta Part A Mol. Biomol. Spectrosc.* **2018**, *201*, 207–215. [CrossRef]
16. Sáez-Hernández, R.; Antela, K.U.; Gallello, G.; Cervera, M.L.; Mauri-Aucejo, A.R. A smartphone-based innovative approach to discriminate red pigments in roman frescoes mock-ups. *J. Cult. Herit.* **2022**, *58*, 156–166. [CrossRef]
17. Merello, P.; Beltrán, P.; García-Diego, F.J. Quantitative Non-Invasive Method for Damage Evaluation in Frescoes: Ariadne's House (Pompeii, Italy). *Environ. Earth Sci.* **2016**, *75*, 165. [CrossRef]
18. Fioretti, G.; Campobasso, C.; Capotorto, S. Digital Photogrammetry as Tool for Mensiochronological Analysis: The Case of St. Maria Veterana Archaeological Site (Triggiano, Italy). *Digit. Appl. Archaeol. Cult. Herit.* **2020**, *19*, e00158. [CrossRef]
19. Horgnies, M.; Bayle, M.; Gueit, E.; Darque-Ceretti, E.; Aucouturier, M. Microstructure and Surface Properties of Frescoes Based on Lime and Cement: The Influence of the Artist's Technique. *Archaeometry* **2015**, *57*, 344–361. [CrossRef]
20. Helvaci, Y.Z.; Dias, L.; Manhita, A.; Martins, S.; Cardoso, A.; Candeias, A.; Gil, M. Tracking Old and New Colours: Material Study of 16th Century Mural Paintings from Évora Cathedral (Southern Portugal). *Color Res. Appl.* **2016**, *41*, 276–282. [CrossRef]
21. Vasco, G.; Serra, A.; Manno, D.; Buccolieri, G.; Calcagnile, L.; Buccolieri, A. Investigations of Byzantine Wall Paintings in the Abbey of Santa Maria Di Cerrate (Italy) in View of Their Restoration. *Spectrochim. Acta Part A Mol. Biomol. Spectrosc.* **2020**, *239*, 118557. [CrossRef]
22. Graziano, S.F.; Rispoli, C.; Guarino, V.; Balassone, G.; Di Maio, G.; Pappalardo, L.; Cappelletti, P.; Damato, G.; De Bonis, A.; Di Benedetto, C.; et al. The Roman Villa of Positano (Campania Region, Southern Italy): Plasters, Tiles and Geoarchaeological Reconstruction. *Int. J. Conserv. Sci.* **2020**, *11*, 319–344.
23. De Benedetto, G.E.; Savino, A.; Fico, D.; Rizzo, D.; Pennetta, A.; Cassiano, A.; Minerva, B. A Multi-Analytical Approach for the Characterisation of the Oldest Pictorial Cycle in the 12th Century Monastery Santa Maria Delle Cerrate. *Open J. Archaeom.* **2013**, *1*, e12. [CrossRef]
24. Bersani, D.; Berzioli, M.; Caglio, S.; Casoli, A.; Lottici, P.P.; Medeghini, L.; Poldi, G.; Zannini, P. An Integrated Multi-Analytical Approach to the Study of the Dome Wall Paintings by Correggio in Parma Cathedral. *Microchem. J.* **2014**, *114*, 80–88. [CrossRef]
25. Romani, M.; Capobianco, G.; Pronti, L.; Colao, F.; Seccaroni, C.; Puiu, A.; Felici, A.C.; Verona-Rinati, G.; Cestelli-Guidi, M.; Tognacci, A.; et al. Analytical Chemistry Approach in Cultural Heritage: The Case of Vincenzo Pasqualoni's Wall Paintings in S. Nicola in Carcere (Rome). *Microchem. J.* **2020**, *156*, 104920. [CrossRef]
26. Alberghina, M.F.; Schiavone, S.; Greco, C.; Saladino, M.L.; Armetta, F.; Renda, V.; Caponetti, E. How Many Secret Details Could a Systematic Multi-Analytical Study Reveal about the Mysterious Fresco Trionfo Della Morte? *Heritage* **2019**, *2*, 145. [CrossRef]
27. Crupi, V.; Fazio, B.; Fiocco, G.; Galli, G.; La Russa, M.F.; Licchelli, M.; Majolino, D.; Malagodi, M.; Ricca, M.; Ruffolo, S.A.; et al. Multi-Analytical Study of Roman Frescoes from Villa Dei Quintili (Rome, Italy). *J. Archaeol. Sci. Rep.* **2018**, *21*, 422–432. [CrossRef]
28. Angelini, I.; Asscher, Y.; Secco, M.; Parisatto, M.; Artioli, G. The Pigments of the Frigidarium in the Sarno Baths, Pompeii: Identification, Stratigraphy and Weathering. *J. Cult. Herit.* **2019**, *40*, 309–316. [CrossRef]
29. Mohanu, I.; Mohanu, D.; Gomoiu, I.; Barbu, O.H.; Fechet, R.M.; Vlad, N.; Voicu, G.; Truşcă, R. Study of the Frescoes in Ioneşti Govorii Wooden Church (Romania) Using Multi-Technique Investigations. *Microchem. J.* **2016**, *126*, 332–340. [CrossRef]
30. Hakobyan, Z.A. The Frescoes of Haghpat Monastery in the Historical-Confessional Context of the 13th Century. *Actual Probl. Theory Hist. Art* **2021**, *11*, 256–267. [CrossRef]
31. Luvidi, L.; Prestileo, F.; De Paoli, M.; Riminesi, C.; Del Fà, R.M.; Magrini, D.; Fratini, F. Diagnostics and Monitoring to Preserve a Hypogeum Site: The Case of the Mithraeum of Marino Laziale (Rome). *Heritage* **2021**, *4*, 235. [CrossRef]
32. Ranalli, G.; Alfano, G.; Belli, C.; Lustrato, G.; Colombini, M.P.; Bonaduce, I.; Zanardini, E.; Abbruscato, P.; Cappitelli, F.; Sorlini, C. Biotechnology applied to cultural heritage: Biorestoration of frescoes using viable bacterial cells and enzymes. *J. Appl. Microbiol.* **2005**, *98*, 73–83. [CrossRef]
33. Galli, A.; Alberghina, M.F.; Re, A.; Magrini, D.; Grifa, C.; Ponterio, R.C.; La Russa, M.F. Special Issue: Results of the II National Research project of AIAr: Archaeometric study of the frescoes by Saturnino Gatti and workshop at the church of San Panfilo in Tornimparte (AQ, Italy). *Appl. Sci.* **2023**. to be submitted.
34. Čiuladienė, A.; Luckutė, A.; Kiuberis, J.; Kareiva, A. Investigation of the Chemical Composition of Red Pigments and Binding Media. *Chemija* **2018**, *29*, 243–256. [CrossRef]
35. Cennini, C. *Il Libro Dell'arte, o Trattato della Pittura di Cennino Cennini*; Milanesi, G., Milanesi, C., Eds.; Le Monnier: Florence, Italy, 1859.
36. Mora, P.; Mora, L.; Philippot, P. *La Conservazione Delle Pitture Murali*; Editrice Compositori: Bologna, Italy, 1999.
37. Piovesan, R.; Mazzoli, C.; Maritan, L.; Cornale, P. Fresco and lime paint: An experimental study and objective criteria for distinguishing between these painting techniques. *Archaeometry* **2012**, *54*, 723–736. [CrossRef]
38. Ergenç, D.; Fort, R.; Varas–Muriel, M.J.; Alvarez de Buergo, M. Mortars and Plasters—How to Characterize Aerial Mortars and Plasters. *Archaeol. Anthr. Sci.* **2021**, *13*, 197. [CrossRef]
39. Bonizzoni, L.; Caglio, S.; Galli, A.; Lanteri, L.; Pelosi, C. Materials and technique: The first look at Saturnino Gatti. *Appl. Sci.* **2023**, *13*, 6842. [CrossRef]

40. Bonizzoni, L.; Caglio, S.; Galli, A.; Germinario, C.; Izzo, F.; Magrini, D. Identifying original and restoration materials through spectroscopic analyses on Saturnino Gatti mural paintings: How far a non-invasive approach can go. *Appl. Sci.* **2023**, *13*, 6638. [CrossRef]
41. Briani, F.; Caridi, F.; Ferella, F.; Gueli, A.M.; Marchegiani, F.; Nisi, S.; Paladini, G.; Pecchioni, E.; Politi, G.; Santo, A.P.; et al. Multi-technique characterization of painting drawings of the pictorial cycle at the San Panfilo Church in Tornimparte (AQ). *Appl. Sci.* **2023**, *13*, 6492. [CrossRef]
42. Armetta, F.; Giuffrida, D.; Ponterio, R.C.; Falcon Martinez, M.F.; Briani, F.; Pecchioni, E.; Santo, A.P.; Ciaramitaro, V.C.; Saladino, M.L. Looking for the original materials and evidence of restoration at the Vault of the San Panfilo Church in Tornimparte (AQ). *Appl. Sci.* **2023**, *13*, 6492. [CrossRef]
43. Andreotti, A.; Izzo, F.C.; Bonaduce, I. Archaeometric study of the mural paintings by Saturnino Gatti and workshop in the Church of San Panfilo—Tornimparte (AQ). The study of organic materials. *Appl. Sci.* **2023**, *13*, 7153. [CrossRef]
44. Centamore, E.; Dramis, F. Note Illustrative Della Carta Geologica d'Italia Alla Scala 1:50.000, Foglio 358-Pescorocchiano. *ISPRA-Serv. Geol. D'Italia* **2010**, 147.
45. Murat, Z. Wall Paintings through the Ages: The Medieval Period (Italy, Twelfth to Fifteenth Century). *Archaeol. Anthr. Sci.* **2021**, *13*, 191. [CrossRef]

Disclaimer/Publisher's Note: The statements, opinions and data contained in all publications are solely those of the individual author(s) and contributor(s) and not of MDPI and/or the editor(s). MDPI and/or the editor(s) disclaim responsibility for any injury to people or property resulting from any ideas, methods, instructions or products referred to in the content.

Article

Characterization of Soluble Salts on the Frescoes by Saturnino Gatti in the Church of San Panfilo in Villagrande di Tornimparte (L'Aquila)

Valeria Comite [1,*], Andrea Bergomi [1,*], Chiara Andrea Lombardi [1,2], Mattia Borelli [1] and Paola Fermo [1]

1 Department of Chemistry, University of Milan, 20133 Milan, Italy; chiara.lombardi@unimi.it (C.A.L.); mattia.borelli@unimi.it (M.B.); paola.fermo@unimi.it (P.F.)
2 Department of Sciences of Antiquity, Sapienza University, 00185 Rome, Italy
* Correspondence: valeria.comite@unimi.it (V.C.); andrea.bergomi@unimi.it (A.B.)

Abstract: Salt crystallization is one of the most dangerous forms of degradation affecting frescoes. This phenomenon can lead to cracking, flaking and detachment of the pictorial layer, ultimately ruining the work of art. However, the characterization of soluble salts via chemical analysis can be employed to determine the conditions of the artifact and establish the proper restoration and/or conservation strategies to be adopted. In this archaeometric study, a first-ever characterization of the soluble salts and related degradation phenomena on the frescoes by Saturnino Gatti in the church of San Panfilo in Villagrande di Tornimparte (L'Aquila) was carried out. Sampling was performed in areas with evident detachments, exfoliations and saline crystallization (efflorescences). Eleven samples of powder and fragments were taken from different panels of the fresco: nine were taken from Panels A, C, D and E and two from the top part of the vault. Chemical characterizations were performed using two analytical techniques: ion chromatography (IC) and attenuated total reflectance Fourier-transform infrared spectroscopy (ATR-FTIR). Ion chromatography was used for the quantification of the main ions and to evaluate the presence of soluble salts, whereas infrared spectroscopy was used to characterize the mineralogical phases. The results show efflorescence consisting of newly formed gypsum and carbonate in samples taken closer to the ground. Furthermore, a good correlation between sodium and chloride ions was observed, indicating the presence of an efflorescence composed of newly formed sodium chloride. Capillary rise and infiltration were highlighted as the main sources of soluble salts. This information will be crucial in guiding future restoration or conservation operations.

Keywords: salts; efflorescence; ion chromatography (IC); infrared spectroscopy

Citation: Comite, V.; Bergomi, A.; Lombardi, C.A.; Borelli, M.; Fermo, P. Characterization of Soluble Salts on the Frescoes by Saturnino Gatti in the Church of San Panfilo in Villagrande di Tornimparte (L'Aquila). *Appl. Sci.* **2023**, *13*, 6623. https://doi.org/10.3390/app13116623

Academic Editor: Asterios Bakolas

Received: 30 April 2023
Revised: 24 May 2023
Accepted: 28 May 2023
Published: 30 May 2023

Copyright: © 2023 by the authors. Licensee MDPI, Basel, Switzerland. This article is an open access article distributed under the terms and conditions of the Creative Commons Attribution (CC BY) license (https://creativecommons.org/licenses/by/4.0/).

1. Introduction

This paper contributes to the Special Issue "Results of the II National Research project of AIAr: archaeometric study of the frescoes by Saturnino Gatti and workshop at the church of San Panfilo in Tornimparte (AQ, Italy)" in which the scientific results of II National Research Project conducted by members of the Italian Association of Archaeometry (AIAr) are discussed and collected. For in-depth details on the aims of the project, see the introduction of the Special Issue [1].

One of the most serious forms of damage affecting porous stone materials used in monuments and other construction materials is directly linked to the physical action exerted by the crystallization of soluble salts [2–6]. This process can cause several degradation phenomena, including flaking, cracks, detachment or staining of the pigment layer, efflorescence, sub-efflorescence and formation of crusts, all of which can lead to irreversible damage of the original material [7,8]. As documented in numerous studies [7,9,10], walls and frescoes suffer greatly from decay induced by salt crystallization, and conservation strategies often need to be put in place in order to salvage the artworks. However, these efforts may in turn exacerbate the problem by introducing additional salts or solutions, as explained in the following paragraph, which may cause the artifacts to decay more rapidly instead of preventing this process.

The salts responsible for this damage can originate from the building material itself or derive from other sources [11]. For instance, they can be channeled from the groundwater and penetrate the wall via capillary rise, with an efficiency that depends on the distribution of the pore size and the wetting behavior of the material [12]. Alternately, they can derive from polluted atmospheres rich in particulate matter, marine aerosols and other polluting agents which are deposited and eventually penetrate the walls [13]. Additionally, acidic and alkaline products used in cleaning and restoration treatments usually contain salts and, if not applied correctly, can give rise to the problems associated with salt crystallization mentioned previously. Finally, they can derive from the metabolic activity of microbes [14]. When the saline solutions that permeate the material, following the evaporation of water, reach suitable thermodynamic conditions, nucleation and growth of crystals can take place. As the water evaporates, the concentration increases and the salt begins to precipitate. The degree of damage is directly proportional to the concentration of the salt and the size of the crystal, and inversely proportional to the size of the pores [12].

Degradation from salt formation derives from the pressure exerted by the crystallization process on the pores of the material. This can occur via different mechanisms: hydrostatic pressure, hydration pressure and linear pressure. Hydrostatic pressure develops when the solution occupies a volume which is smaller than the one of the precipitating crystals [15]. In some cases, this may only constitute a minor effect, since the increase in pressure is associated with an increase in solubility, limiting the impact of salt crystallization. Hydration pressure occurs following the increase in volume generated by the hydration of the salt [16]. This effect is strictly related to the nature and physical–chemical characteristics of the salt; therefore, the impact may be different from case to case. Finally, linear pressure develops following the crystallization of a salt in the pores of the material, and the exerted pressure will be inversely proportional to the pore size [17]. Many calcarenites with pore sizes smaller than 0.1 micron suffer greatly from this type of degradation [15,16]. Often, these three mechanisms have a combined effect, leading to an increase in pore size and eventually to cracks, efflorescence and sub-efflorescence [16]. Due to the combined effect of all these mechanisms, the pore increases in size and eventually can lead to cracks, efflorescence and/or sub-efflorescence.

According to Zehnder et. al. (1991), the nature of the salts that can be formed varies along the vertical profile of the wall, depending on their solubility. Typically, the relatively less soluble salts, such as calcium carbonate and gypsum, can precipitate more easily and, therefore, are found in the areas closer to the ground. Instead, sulphate and nitrate salts will precipitate in the intermediate sections. Finally, the most soluble and deliquescent salts are found in the upper parts of the wall; these include nitrates and chlorides of sodium and magnesium, which can concentrate and accumulate for centuries. The intermediate sections are usually the most degraded parts because of the favorable conditions for the formation of efflorescence and sub-efflorescence [12]. Indeed, in the areas closer to the ground, the rate of capillary rise of the water is faster than the rate of evaporation from the walls, therefore inhibiting supersaturation of the solution and crystal formation. Instead, as the rate of capillary rise decreases along the wall, the solution becomes supersaturated, and efflorescence starts to form on the outside. Finally, when the rate of rise decreases below the rate of evaporation, the salts are formed inside the walls (sub-efflorescence).

The present study is concerned with the characterization of soluble salts on the frescoes by Saturnino Gatti in the Church of San Panfilo in Villagrande di Tornimparte, which is located in the city of L'Aquila, in the Abruzzo region of Italy. Eleven samples were retrieved at different heights along the walls of the church hosting the frescoes and characterized using ion chromatography (IC) and attenuated total reflectance Fourier-transform infrared spectroscopy (ATR-FTIR). The former was used to determine the ionic composition, in terms of both anions and cations, whereas the latter was employed to identify the most abundant mineralogical phases. The aim of this research is to characterize the salts in the masonry that caused the degradation observed on the surfaces of the church, both in terms of nature and concentration. A complete characterization of the fresco was performed by sampling areas at different heights, covering the entire structure. Thanks to the chemical analysis, it was possible to trace the sources of the soluble salts. The results of this study will also be useful when conducting the restoration processes of the walls.

2. Materials and Methods

2.1. Sampling

Eleven samples of powder and fragments were taken from different panels of the fresco in order to characterize the degradation produced by the presence of salts. The samples (Figure 1 and Table 1) were taken from areas where salt efflorescence and detachment of the plaster surface were visible. Sampling was carried out at different heights: seven fragments were taken close to the ground level (SG_1 to 7), two samples were taken near window splays (SG_8 and SG_29), and two samples were taken from the top of the vault (SG_37A and SG_41B).

Figure 1. Sampling points of **Panels A,C–E** and the upper part of the vault on the walls of the Church of San Panfilo in Villagrande di Tornimparte (L'Aquila). The images were obtained with the photogrammetric survey.

Table 1. Sampling details: sampling area, sample ID, description, and type of sample.

Sampling Area	Sample ID	Description	Type of Sample
A	SG_6	Original area (pictorial layer and plaster) affected by detachments and fractures, with efflorescence.	Powders and fragments
A	SG_7	Integration area affected by partial detachment of the pictorial film. Exfoliations can be observed due to the presence of saline efflorescence.	Powders
A	SG_8	Original area (dark yellow pictorial layer and plaster), detached, affected by saline efflorescence, not cohesive, and fragile.	Powders and fragments
C	SG_4	Integration area with yellow-brown pictorial finishing, affected by detachments and exfoliations—presence of salt efflorescence.	Powders and fragments
C	SG_5	Original area (pictorial layer and plaster) affected by detachments and fractures.	Fragments of various dimensions
C	SG_29	Saline efflorescence selectively retrieved from the surface.	Powders and plaster residues
D	SG_3	Integration area without pictorial finishing and affected by detachment and exfoliation—presence of salt efflorescence.	Powders
E	SG_1	Black paint layer and plaster layer; area affected by substrate detachment and efflorescence.	Fragments of various dimensions
E	SG_2	Plaster layer; area affected by complete detachment of the pictorial film and efflorescence.	Powders (sampled up to 1 cm from the surface)
Vault, summit	SG_37A	Grey pictorial layer on degraded plaster, retrieved from degraded area with detachments of the superficial layers.	Powders and fragments
Vault, summit	SG_41B	Purple pictorial layer on degraded plaster, retrieved from degraded area with detachments of the superficial layers.	Powders and fragments

2.2. Ion Chromatography Analysis

Quantification of the main ionic components and evaluation of the presence of soluble salts were performed with ion chromatography (IC). Anions (NO_3^-, SO_4^{2-}, Cl^-) and cations (Na^+, K^+, Ca^{2+}, Mg^{2+}, NH_4^+) were determined using an HPLC Dionex ICS-1000 Ion Chromatography System (Thermo Scientific, Dionex Corporation, Sunnyvale, CA, USA) equipped with a conductivity detector. Anion analysis was carried out with an Ion Pac AS14A IC Column, using 8 mM Na_2CO_3/1 mM $NaHCO_3$ as the eluent, isocratic elution, a constant flux of 1.5 mL min^{-1} and an ULTRA anionic self-healing suppression (ASRS-ULTRA), whereas cations were analyzed with an Ion Pac CS12A IC Column, using 20 mM methanesulfonic acid (MSA) as the eluent, isocratic elution, a constant flux of 1.5 mL min^{-1} and an ULTRA cationic self-healing suppression (CSRS-ULTRA).

In order to prepare the solutions to be analyzed, a small portion of the sample was ground in an agate mortar. Of this powder, around 2 mg was transferred in a plastic test-tube and treated with 10 mL of milli-q water (Merck Millipore Milli-Q, Burlington, MA, USA). The suspensions were immersed in an ultrasonic bath for 1 h, centrifuged at 3000 rpm for 3 min and filtered using 0.45 µm non-sterile hydrophilic membranes (PTFE Millex-14 LCR, 25 mm, Millex® Syringe Filters, Merck Millipore, Burlington, MA, USA) before injection in the instrument.

2.3. ATR-FTIR Analysis

Fourier-transform infrared spectroscopy was carried out in attenuated total reflectance mode (ATR-FTIR) in order to identify the main components and phases of the samples. The instrumentation used to perform the analyses was a Nicolet 380 FTIR spectrometer (Thermo Electron Corporation, Waltham, MA, USA). The detection window used was between 400 cm^{-1} and 4000 cm^{-1}, and 64 scans with a resolution of 4 cm^{-1} were performed, along with smoothing operations (15 points).

3. Results

3.1. Ion Chromatography Analysis

The results of the determination of the main ionic species are shown in Table 2.

Table 2. Concentrations of cations (sodium, ammonium, potassium, magnesium and calcium) and anions (chloride, nitrate and sulphate) in ppm (µg g^{-1}) in the analyzed samples.

Sampling Area	Samples ID	Cations					Anions		
		Sodium	Ammonium	Potassium	Magnesium	Calcium	Chloride	Nitrate	Sulphate
A	SG_6	5.98	n.d.	1.32	2.08	326.23	7.78	65.21	730.11
	SG_7	2.01	0.17	0.35	0.93	350.20	3.13	24.84	843.13
	SG_8	3.93	0.40	2.86	1.67	58.32	7.02	23.09	16.68
C	SG_4	1.05	1.09	n.d.	0.67	369.36	1.36	14.15	922.52
	SG_5	1.83	n.d.	0.53	2.22	97.77	2.76	16.91	123.47
	SG_29	1.94	n.d.	1.01	n.d.	121.07	1.62	10.05	284.24
D	SG_3	2.54	n.d.	n.d.	0.71	483.05	4.15	6.12	1256.41
E	SG_1	9.02	0.48	6.34	6.11	601.09	9.12	198.78	1475.08
	SG_2	5.22	0.40	2.51	2.83	68.80	10.71	74.89	12.14
Vault summit	SG_37A	3.13	2.20	8.90	n.d.	3.86	4.01	14.20	27.56
	SG_41B	2.37	1.11	0.44	2.58	60.84	5.06	11.86	26.72

n.d. = not detected. Standard error = 5%.

The concentrations of anions (Table 2) varied between the different samples; even between those taken from the same panel, but at different heights. For almost all the samples, sulphates were the species present in highest concentration, with higher values (>700 ppm) found in samples collected near the ground: SG_6 and 7 (Panel A), SG_4 (Panel C), SG_3 (Panel D), SG_1 (Panel E). Overall, it is possible to appreciate a relationship between sulphate concentration and sampling height. Indeed, the samples taken close to the window splays or in the vault showed, on average, lower concentrations of sulphates compared to samples taken at lower heights. A clear trend in terms of sampling height could not be observed for chlorides and nitrates. In this case, the differences between the samples are related to sampling site (panel), with Panels A and E showing the highest average concentrations of both species.

Moving on to cations, almost all the samples showed high concentrations of calcium ions (above 50 ppm, except for SG_37A). A correlation can be observed between the concentrations of calcium and sulphate ions. Indeed, as was the case for sulphates, it is possible to observe, on average, a relationship with sampling height, with calcium content being lower for samples taken at greater heights. All other cations were found in concentrations below 10 ppm, therefore secondary with respect to calcium. In this case, no significant correlations between concentration values and sampling height or sampling site (panel) could be appreciated.

3.2. ATR-FTIR Analysis

The samples were also analyzed using attenuated total reflectance Fourier-transform infrared spectroscopy (ATR-FTIR). Table 3 summarizes the characteristic bands observed in the different samples. For most of the samples (SG_1, SG_3, SG_4, SG_5; SG_6, SG_7 and SG_29), typical gypsum bands were observed, in agreement with the IC results. Indeed, these samples showed high concentrations of sulphates (above 100 ppm), whereas samples SG_2, SG_8, SG_37A and SG_41B were associated with sulphate concentrations below 30 ppm. Based on the combined results of both IC and ATR-FTIR techniques, it is possible to assume that most of the ionic sulphate concentration derives from gypsum formation. Moreover, the characteristic calcite bands were recognized in all samples, which explains the nature of the high calcium concentrations observed in IC analysis. Finally, samples SG_2,

SG_7, SG_8, SG_29, SG_37A and SG_41B also showed silicate bands, probably attributable to the substrate.

Table 3. ATR-FTIR bands in the analyzed samples (cm^{-1}).

Sampling Area	Samples ID	Silicates	Carbonates	Sulphates (Gypsum)
A	SG_6		1412, 872	3523, 3400, 1683, 1620, 1109, 667, 598, 420
	SG_7			3522, 3400, 1683, 1620, 1107, 667, 597, 418
	SG_8	1012, 777, 459	1405, 872, 712	
C	SG_4		1426, 873	3525, 3401, 1683, 1620, 1110, 668, 598, 419
	SG_5		1421, 873	3535, 3401, 1683, 1620, 1109, 668, 599, 454
	SG_29		1413, 872	3235, 1643, 1110, 598
D	SG_3		1429, 874	3517, 3400, 1683, 1619, 1105, 667, 597, 418
E	SG_1		1410, 872, 712	3522, 3400, 1683, 1620, 1108, 667, 597, 418
	SG_2	1011, 777, 458	1407, 872, 711	
Vault summit	SG_37A	1005, 778, 459	1411, 873, 712	
	SG_41B	1007, 777, 460	1410, 872, 712	

4. Discussion

The investigations carried out on the samples made it possible to characterize the different phases of salt crystallization in the walls of the Church of San Panfilo in Villagrande di Tornimparte. These were often related to the sampling height. General graphs of samples from the same area taken at different heights are shown below. It is interesting to observe how some samples, e.g., SG_1 and SG_2 (Figure 2) taken from Panel E, despite coming from the same area, near the ground, show different salt crystallization conditions.

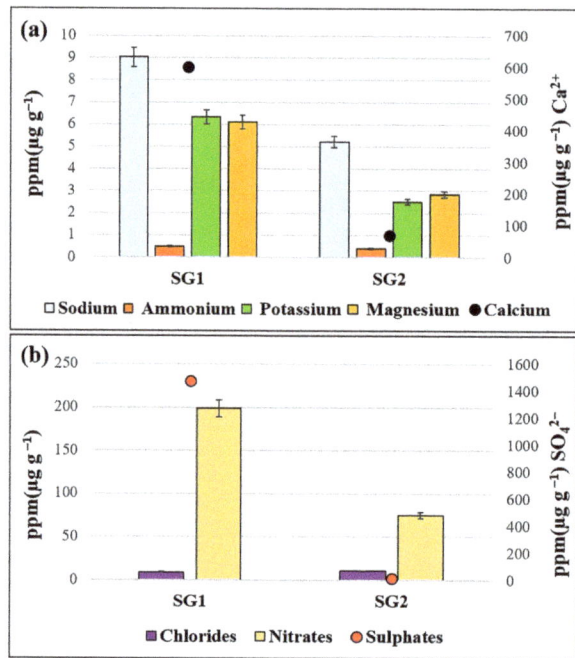

Figure 2. Ion concentration of samples SG_1 and SG_2 taken from Panel E. (**a**) Ppm (µg g^{-1}) concentrations of sodium, ammonium, potassium, magnesium and calcium cations; (**b**) ppm concentrations (µg g^{-1}) of chloride, nitrate and sulphate ions.

In fact, sample SG_1 shows a good concentration trend between calcium and sulphate ions and lower concentrations of the other ions (Figure 2). This observation allowed us to hypothesize that the area under investigation is affected by the sulphation process, which leads to the formation of newly formed crystalline gypsum in the form of efflorescence. This hypothesis is confirmed when observing the ATR-FTIR spectrum acquired for this sample (Figure 3, Table 3). The characteristic peaks of the gypsum centered at 3522, 3400, 1683, 1620, 1108, 667, 597 and 418 cm^{-1} can all be observed [18,19]. In addition, O–H stretching and bending vibrations of calcium carbonate were found, with peaks at 1412, 872 and 712 cm^{-1} [19,20]. These peaks can also be attributable to newly formed carbonate salts and not to the substrate, since the samples were taken at salt efflorescence.

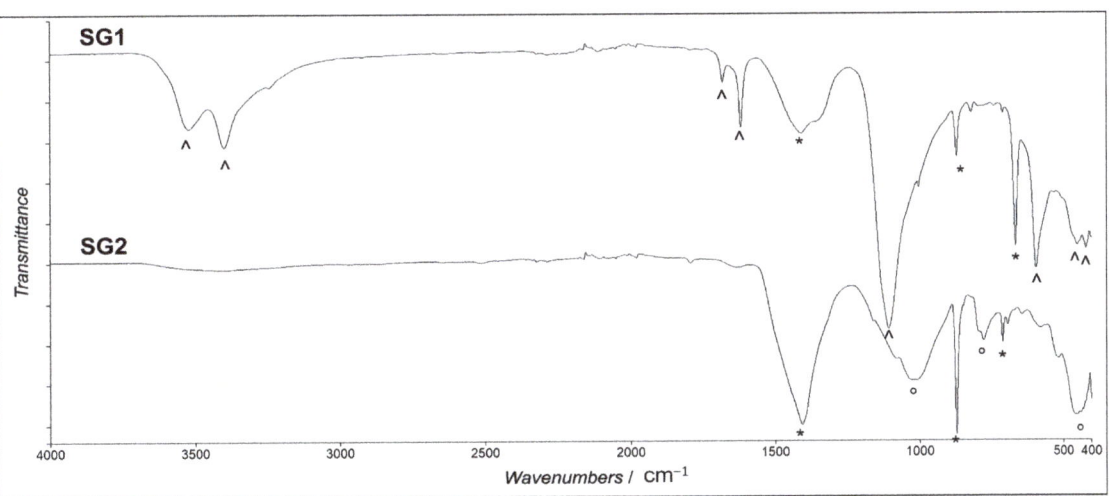

Figure 3. ATR-FTIR spectra of samples SG_1 and SG_2 taken from Panel E. Marker bands of carbonates (*), silicates (°) and sulphates (^) are highlighted.

In contrast, sample SG_2 shows low concentrations of these ions; in particular, sulphate ions (Figure 2). In fact, the typical bands of gypsum were not observed (Figure 3, Table 3). Sample SG_2, on the other hand, shows the classic carbonate peaks (Figure 3, Table 3), which can be attributable to efflorescence produced by carbonate salts. Indeed, it is known in the literature [14] that the area closer to the ground is where the relatively less soluble salts, such as calcium carbonate and gypsum, tend to form and precipitate as newly formed crystals, generating salt efflorescence which causes the observed detachment of the painted surface. The tensions generated by the increase in size of the salt crystals in the pores of the material can cause the detachment of the most superficial parts of the plaster and/or colored finishes, or, in the most severe cases, can even damage part of the load-bearing masonry. As was demonstrated by Kilian et.al. (2023), gypsum accumulation can cause hardening of the surface, which tends to decrease the superficial porosity of the material. The formation of such crusts reduces water evaporation from the surface and consequently increases sub-efflorescence formation [21].

Moving on to Panel A, the samples taken from this sampling site at different heights show different concentrations of calcium and sulphate ions (Figure 4).

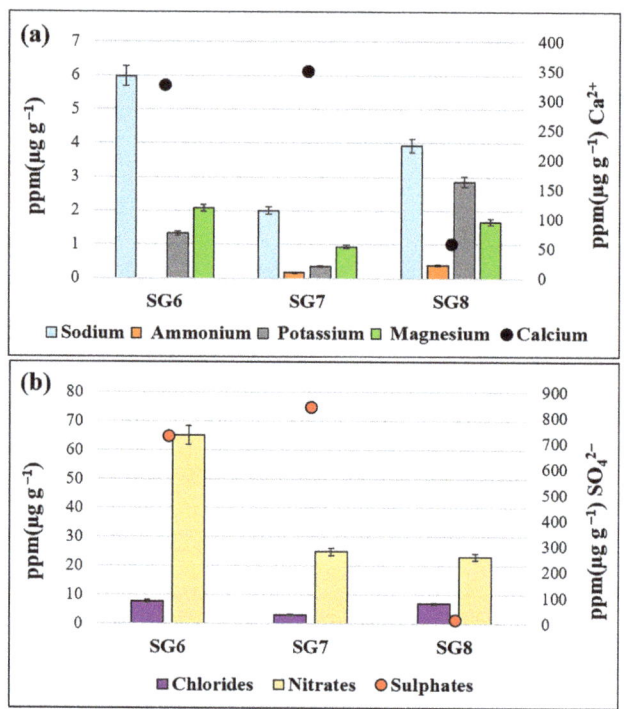

Figure 4. Ion concentration of samples SG_6, SG_7 and SG_8 taken from Panel A. (**a**) Ppm (µg g^{-1}) concentrations of sodium, ammonium, potassium, magnesium and calcium cations; (**b**) ppm concentrations (µg g^{-1}) of chloride, nitrate and sulphate anions.

In fact, samples SG_6 and SG_7, taken a few cm above ground level, show higher values of these ions than sample SG_8, which was taken at a greater height close to a window splay (Figure 4). This result indicates that the wall of Panel A close to the ground level are affected by efflorescence produced by sulphation, which produces newly formed gypsum crystals that lead to the detachment of the painted surface [22]. For these samples, the presence of gypsum is confirmed by infrared analysis (Figure 5, Table 3). Typical calcite bands were observed (Figure 5, Table 3) [19,20], and can be attributed to carbonate salts, which contribute to the degradation of the wall.

A similar result was obtained by analyzing samples taken from Panel C. Samples SG_4 and SG_5 were taken at lower heights, near the ground level, whereas sample SG_29 was retrieved at a greater height, near the window splay (Figures 6 and 7, Table 3). In this case, sample SG_4 shows significantly higher concentrations of both calcium and sulphate ions, with respect to samples SG_5 and SG_29. These results indicate an even stronger relationship between sampling height and ionic concentration (calcium and sulphate), once again reaffirming the previous conclusions that gypsum neoformation salts deriving from capillary rise tend to be located closer to the ground, in the vicinity of the walking surface. The pressure exerted by these salts causes the rupture of the porous material, leading to the degradation of the walls and, therefore, serious damage to the work of art, in this case the fresco. Indeed, the sampling site from which sample SG_4 was retrieved appears highly degraded, and most of the pictorial layer is no longer visible (Figure 1). In contrast, the area from which sample SG_5 was taken shows signs of superficial degradation, but the pictorial layer is still intact. Compared to the other samples taken close to the window splays (SG_8) or from the vault (SG_37A, SG_41B), sample SG_29 shows relatively high concentrations of calcium and sulphate. This may be attributable to external infiltrations that are specific to this sampling site.

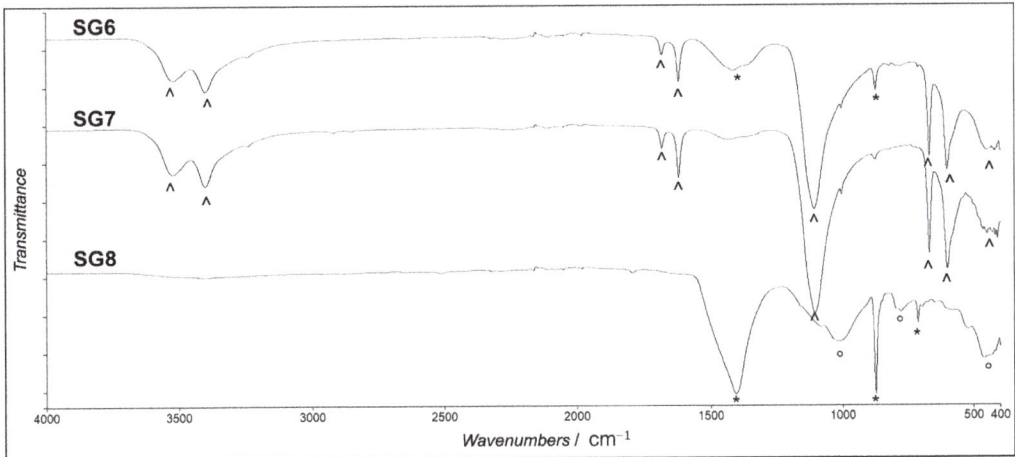

Figure 5. ATR-FTIR spectra of samples SG_6, SG_7 and SG_8 taken from Panel A. Marker bands of carbonates (*), silicates (°) and sulphates (^) are highlighted.

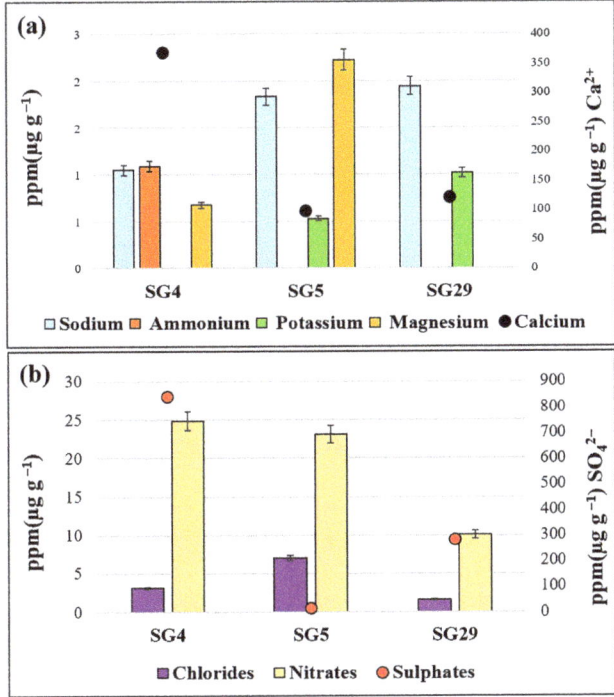

Figure 6. Ion concentrations of samples SG_4, SG_5 and SG_29 taken from Panel C. (**a**) Ppm (µg g^{-1}) concentrations of sodium, ammonium, potassium, magnesium and calcium cations; (**b**) ppm concentrations (µg g^{-1}) of chloride, nitrate and sulphate anions.

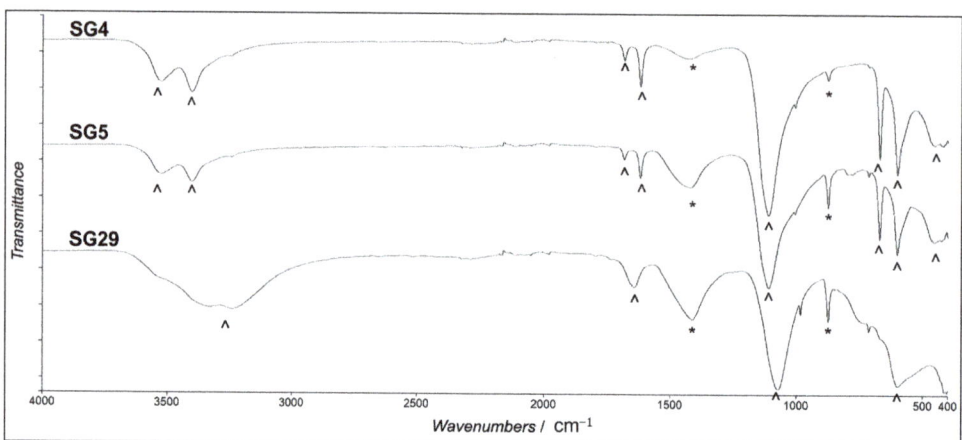

Figure 7. ATR-FTIR spectra of samples SG_4, SG_5 and SG_229 taken from Panel C. Marker bands of carbonates (*) and sulphates (^) are highlighted.

The sample taken from Panel D (SG_3) also showed high concentrations of sulphate and calcium ions (Figure 8a), attributable to the same degradation phenomena, which was also confirmed by observing the spectrum obtained by infrared spectroscopy (Figure 8b, Table 3). In fact, the sampled area is characterized by the complete loss of the plaster and of the pictorial layer caused by the crystallization cycles of these salts (Figure 1).

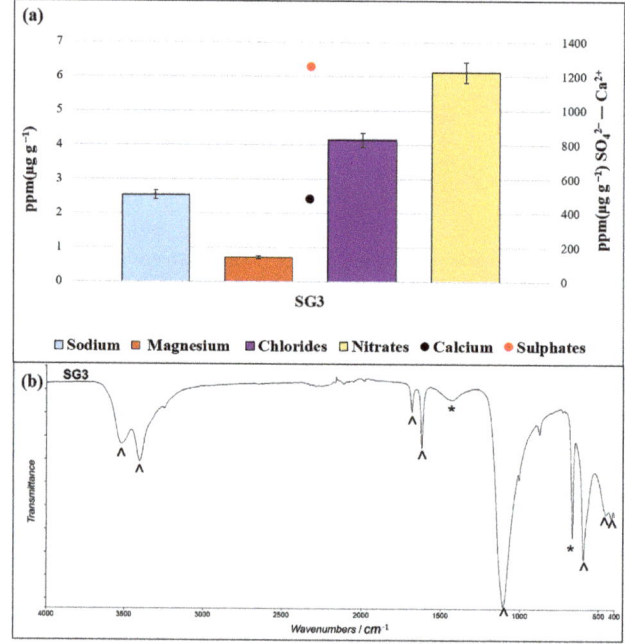

Figure 8. Sample SG3 taken from Panel D. (**a**) Ion concentrations of sodium, magnesium, chlorides, nitrates, calcium, sulphate in ppm (μg g^{-1}); (**b**) ATR-FTIR spectrum: marker bands of carbonates (*) and sulphates (^) are highlighted.

Finally, samples taken at the top of the vault do not show high values of sulphate and calcium ions (moderate concentrations were only obtained in sample SG_37A) (Figure 9).

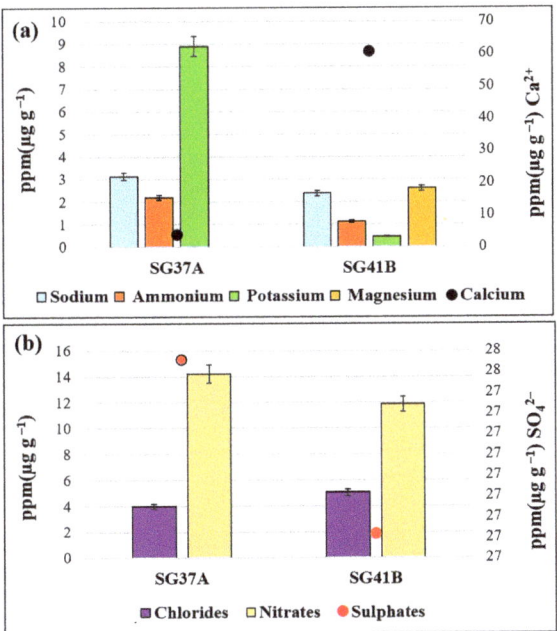

Figure 9. Ion concentration of samples SG_37A and SG_41B taken from the top of the vaults. (**a**) Ppm (µg g^{-1}) concentrations of sodium, ammonium, potassium, magnesium and calcium cations; (**b**) ppm concentrations (µg g^{-1}) of chloride, nitrate and sulphate anions.

Based on the results obtained, it is difficult to assume the precipitation of a specific salt. In fact, from the infrared spectra of samples taken from the detached surface (Figure 10, Table 3), carbonate bands were observed [19,20] that can be attributed either to newly formed carbonate crystals or to the paint substrate.

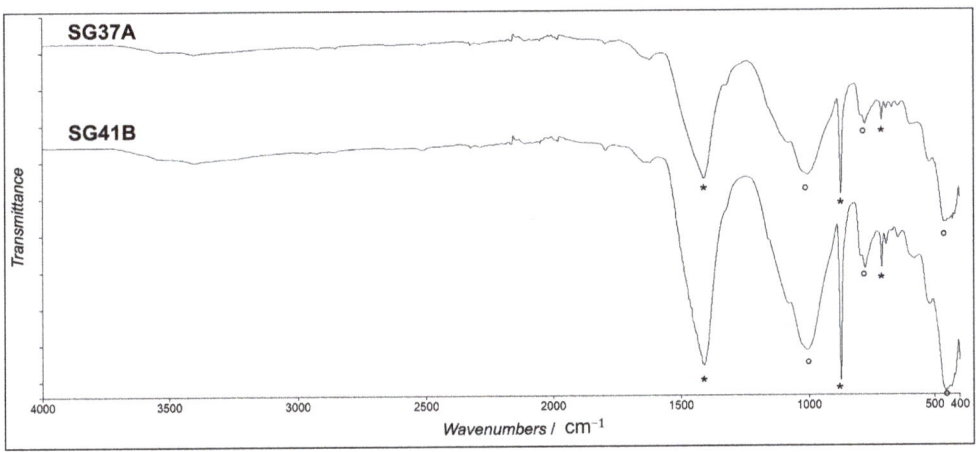

Figure 10. ATR-FTIR spectra of samples SG_37A and SG_41B taken from the top of the vaults. Marker bands of carbonates (*) and silicates (°) are highlighted.

The graph shown in Figure 11a indicates that the calcium and sulphur trends are perfectly aligned for many of the analyzed samples in which efflorescence produced by poorly soluble gypsum salts was assumed. Gypsum has a low solubility and hygroscopicity compared to most salts, and its deliquescence occurs at a very high relative humidity (over 99.9%, according to [12]). Hence, it is stable in a variety of environments, even extremely humid ones. It should be noted that studies in literature [12] show that the decay rate of wall paintings due to salt contamination originating from gypsum crystallization causes significant deterioration, but also that it is 10 to 100 times slower than that caused by more soluble salts. In fact, the areas where newly formed gypsum was found did show efflorescence, but detachment of the paint surface was not always very evident (except for the area where SG_4 and SG_3 was sampled). This is supported by other studies which indicate that the deterioration caused by gypsum in stone and similar building materials is not produced by hydration pressures [12], but is the result of its crystallization within the porous matrix of the material, as with any other non-hydrating salt. Because of these characteristics, once gypsum is deposited in the pores of a building material or on a pictorial surface, such as a fresco, it will tend to accumulate over time. Its extremely low mobility also explains why desalination of materials contaminated with calcium sulphate is very difficult [12,23,24].

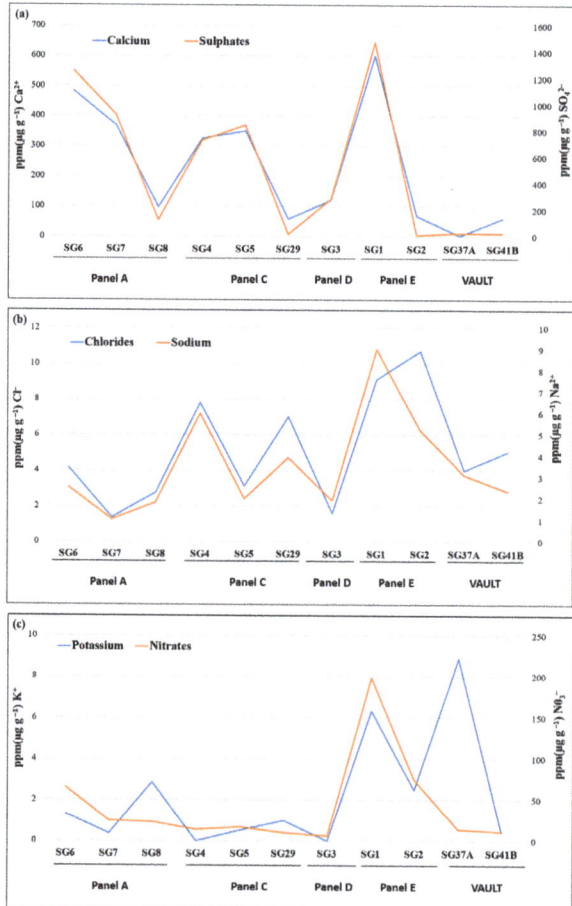

Figure 11. Trend of the concentrations ($\mu g\ g^{-1}$) of certain ions observed in the samples analyzed. (**a**) Calcium and sulphate; (**b**) chloride and sodium; (**c**) potassium and nitrates.

Furthermore, a comparison was made between the concentrations of sodium and chloride ions (Figure 11b), and a good correlation was observed for samples taken from Panel A (SG_6, SG_7, SG_8), Panel C (except for sample SG_29) and the samples taken from the top of the vault (SG_37A). This suggests that sodium chloride salts, such as halite, are present in these areas in addition to newly formed gypsum. Finally, the samples that showed correlations between potassium and nitrate ions (Figure 11c) were those taken from Panel A (SG_6 and SG_7, with the exception of sample SG_8), the samples taken from Panel C (SG_5 and SG_29, with the exception of sample SG_4) and the samples taken from Panel E (SG_1 and SG_2, which show the highest concentrations of these species). For these samples, the presence of salts such as niter is assumed. Moreover, chloride and nitrate salts (particularly with sodium and magnesium) are more soluble and deliquescent than others and do not usually precipitate on the surface of the wall [25]. However, when the outdoor climate becomes exceptionally dry, efflorescence of this nature can be observed on surfaces. This means that these salts accumulate locally over prolonged periods of time, which can be as long as several decades, and tend to keep the masonry moist at all times, except in very dry conditions [12]. The types of decay that form on masonry usually result in swelling and detachment of the plaster, as was observed on the panels of the walls studied. The distribution of salts within a wall also depends on the actual mixture of salts present and their origin. As discussed by Steiger [26], based on the in-depth analysis of the north façade of a convent in northern Bavaria and other monuments [12], the presence of nitrate, potassium, and magnesium, as well as chloride and sodium, depends on the mixture of salts present and their origins, and is the result of rising damp.

Regarding the mechanism of the saline degradation [14], the results obtained in this study allow us to hypothesize that the degradation produced by the salts on the walls of this monument can be attributed mainly to capillary rise. Fractionation during ascending transport is revealed by the shift in the potassium/magnesium (K/Mg) ratio, which decreases with increasing height [12]. In fact, this ratio, calculated based on the data obtained from the different samples, showed a decrease in this parameter with increasing height. In the samples taken from Panel E (SG_1 and SG_2, K/Mg ratio = 1.03; 0.89, respectively), Panel C (SG_4 and SG_5, K/Mg ratio = 0.98; 0.24, respectively) and Panel A (SG_6 and SG_7, K/Mg ratio = 0.63; 0.38, respectively), the K/Mg ratio indicates an ascending transport of salts due to capillary rise from the ground. Finally, for the samples taken from the splays of the windows (SG_8, Panel A; SG_29, Panel C) or from the top of the vaults (SG_37A, SG_41B), it can be assumed that the detachment observed on the surfaces cannot be attributed to capillary rise from the ground, but is probably attributable to infiltration.

5. Conclusions

Thanks to the use of two analytical techniques (IC and ATR-FTIR), it was possible to characterize the degradation due to salt crystallization of the masonry of the fresco by Saturnino Gatti in the church of San Panfilo in Villagrande di Tornimparte (L'Aquila). The results highlighted the different types of efflorescence that probably caused the detachment of the painted surface and of the plaster.

Specifically, areas were identified where efflorescence was mainly caused by newly formed crystals of gypsum and calcium carbonate. These two types of salts were found in almost all the samples analyzed. Higher concentrations of gypsum were present, especially for those samples taken at a lower height, such as Panels A, C, D and E. Efflorescence composed of newly formed sodium chloride, halite, was present in samples SG_4, 5, 6, 7, 8 and 37A, with varying concentrations (higher concentrations for samples SG_6 and 8). This evidence was observed in the samples taken at slightly greater heights, with higher values detected in Panels A and C and in the sample taken from the upper vault (SG_37A). Efflorescence composed of potassium nitrate was present in samples SG_1 (higher concentrations) and SG_2, 6 and 7 (lower concentrations) (Panels E and A).

Finally, the results suggested that the degradation produced by efflorescence was due mainly to the capillary rise from the ground, with the exception of the samples taken close to the window splays and from the vault, in which infiltration could be responsible. The research carried out will allowed us to provide indications in order to perform targeted interventions regarding possible restoration work aimed at desalinizing the church walls.

Author Contributions: Conceptualization, V.C. and P.F.; methodology, A.B., C.A.L. and M.B.; validation, V.C., A.B. and P.F.; formal analysis, C.A.L. and M.B.; investigation, V.C. and A.B.; resources, V.C. and A.B.; data curation, V.C.; writing—original draft preparation, V.C., P.F. and A.B.; writing—review and editing, V.C. and A.B.; supervision, V.C. and P.F.; project administration, P.F. All authors have read and agreed to the published version of the manuscript.

Funding: This work was performed in the framework of the Project Tornimparte—"Archeometric investigation of the pictorial cycle of Saturnino Gatti in Tornimparte (AQ, Italy)" sponsored in 2021 by the Italian Association of Archeometry AIAR (www.associazioneaiar.com, last accessed: 27 May 2023).

Institutional Review Board Statement: Not applicable.

Informed Consent Statement: Not applicable.

Data Availability Statement: Data sharing is not applicable. No new data were created or analyzed in this study. Data sharing is not applicable to this article.

Conflicts of Interest: The authors declare no conflict of interest.

References

1. Galli, A.; Alberghina, M.F.; Re, A.; Magrini, D.; Grifa, C.; Ponterio, R.C.; La Russa, M.F. Special Issue: Results of the II National Research project of AIAr: Archaeometric study of the frescoes by Saturnino Gatti and workshop at the church of San Panfilo in Tornimparte (AQ, Italy). *Appl. Sci.* **2023**. to be submitted.
2. D'Agostino, D.; Congedo, P.M.; Cataldo, R. Computational Fluid Dynamics (CFD) Modeling of Microclimate for Salts Crystallization Control and Artworks Conservation. *J. Cult. Herit.* **2014**, *15*, 448–457. [CrossRef]
3. Flatt, R.J.; Caruso, F.; Sanchez, A.M.A.; Scherer, G.W. Chemo-Mechanics of Salt Damage in Stone. *Nat. Commun.* **2014**, *5*, 4823. [CrossRef]
4. Kotulanova, E.; Schweigstillova, J.; Švarcová, S.; Hradil, D.; Bezdička, P.; Grygar, T. Wall painting damage by salts: Causes and mechanisms. *Acta Res. Rep.* **2009**, *18*, 27–31.
5. D'Altri, A.M.; de Miranda, S.; Beck, K.; De Kock, T.; Derluyn, H. Towards a More Effective and Reliable Salt Crystallisation Test for Porous Building Materials: Predictive Modelling of Sodium Chloride Salt Distribution. *Constr. Build. Mater.* **2021**, *304*, 124436. [CrossRef]
6. Benavente, D.; Cueto, N.; Martinez-Martinez, J.; Garcia del Cura, M.A.; Canaveras, J.C. The influence of petrophyrical properties on the salt weathering of porous building rocks. *Environ. Geol.* **2007**, *52*, 215–224. [CrossRef]
7. Giustetto, R.; Moschella, E.M.; Cristellotti, M.; Costa, E. Deterioration of building materials and artworks in the 'Santa Maria della Stella' chuch, Saluzzo (Italy): Causes of decay and possible remedies. *Stud. Conserv.* **2017**, *62*, 474–493. [CrossRef]
8. Yu, W.; Yang, L.; Zhao, J.; Luo, H. Study on the Visualization of Transport and Crystallization of Salt Solution in Simulated Wall Painting. *Crystals* **2022**, *12*, 351. [CrossRef]
9. Prieto-Taboada, N.; Fdez-Ortiz De Vallejuelo, S.; Veneranda, M.; Marcaida, I.; Morillas, H.; Maguregui, M.; Castro, K.; De Carolis, E.; Osanna, M.; Madariaga, J.M. Study of the Soluble Salts Formation in a Recently Restored House of Pompeii by In-Situ Raman Spectroscopy. *Sci. Rep.* **2018**, *8*, 1613. [CrossRef]
10. Bosch-Roig, P.; Lustrato, G.; Zanardini, E.; Ranalli, G. Biocleaning of Cultural Heritage Stone Surfaces and Frescoes: Which Delivery System Can Be the Most Appropriate? *Ann. Microbiol.* **2015**, *65*, 1227–1241. [CrossRef]
11. Kilian, R.; Borgatta, L.; Wendler, E. Investigation of the Deterioration Mechanisms Induced by Moisture and Soluble Salts in the Necropolis of Porta Nocera, Pompeii (Italy). *Herit. Sci.* **2023**, *11*, 72. [CrossRef]
12. Scherer, G.W. Stress from Crystallization of Salt. *Cem. Concr. Res.* **2004**, *34*, 1613–1624. [CrossRef]
13. Maguregui, M.; Knuutinen, U.; Martínez-Arkarazo, I.; Giakoumaki, A.; Castro, K.; Madariaga, J.M. Field Raman Analysis to Diagnose the Conservation State of Excavated Walls and Wall Paintings in the Archaeological Site of Pompeii (Italy). *J. Raman Spectrosc.* **2012**, *43*, 1747–1753. [CrossRef]
14. Arnold, A.; Zehnder, K. Monitoring Wall Paintings Affected by Soluble Salts. In *The Conservation of Wall Paintings; Proceedings of a Symposium Organized by the Courtauld Institute of Art and the Getty Conservation Institute, London, July 13–16 1987*; Getty Conservation Institute: Los Angeles, CA, USA, 1991; pp. 103–135. ISBN 0-89236-162-X.
15. Correns, C.W. Growth and diffusion of crystals under linear pressure. *Faraday Discuss.* **1949**, *5*, 267. [CrossRef]
16. Winkler, E.M.; Wilhelm, E.J. Salt burst by hydration pressures in architectural stone in urban atmosphere. *GSA Bull.* **1970**, *81*, 567–572. [CrossRef]

17. Everett, D.H. The thermodynamics of frost damage to porous solids. *Trans. Faraday Soc.* **1961**, *57*, 1541. [CrossRef]
18. Vahur, S.; Teearu, A.; Peets, P.; Joosu, L.; Leito, I. ATR-FT-IR Spectral Collection of Conservation Materials in the Extended Region of 4000–80 °C m^{-1}. *Anal. Bioanal. Chem.* **2016**, *408*, 3373–3379. [CrossRef]
19. Rovella, N.; Aly, N.; Comite, V.; Ruffolo, S.A.; Ricca, M.; Fermo, P.; de Buergo, M.A.; Russa, M.F. La A Methodological Approach to Define the State of Conservation of the Stone Materials Used in the Cairo Historical Heritage (Egypt). *Archaeol. Anthr. Sci.* **2020**, *12*, 178. [CrossRef]
20. Chukanov, N.V.; Chervonnyi, A.D. *Infrared Spectroscopy of Minerals and Related Compounds*; Springer: Berlin/Heidelberg, Germany, 2016.
21. Modestou, S.; Theodoridou, M.; Ioannou, I. Micro-Destructive Mapping of the Salt Crystallization Front in Limestone. *Eng. Geol.* **2015**, *193*, 337–347. [CrossRef]
22. Charola, A.E.; Wendler, E. An Overview of the Water-Porous Building Materials Interactions. *Restor. Build. Monum.* **2016**, *21*, 55–65. [CrossRef]
23. Charola, A.E.; Pühringer, J.; Steiger, M. Gypsum: A Review of Its Role in the Deterioration of Building Materials. *Environ. Geol.* **2007**, *52*, 339–352. [CrossRef]
24. Chwast, J.; Todorović, J.; Janssen, H.; Elsen, J. Gypsum Efflorescence on Clay Brick Masonry: Field Survey and Literature Study. *Constr. Build. Mater.* **2015**, *85*, 57–64. [CrossRef]
25. Ju, X.; Feng, W.; Zhang, Y.; Zhao, H. Stress from Crystllization in Ideal Pores. *Yanshilixue Yu Gongcheng Xuebao/Chin. J. Rock. Mech. Eng.* **2016**, *35*, 2787–2794. [CrossRef]
26. Steiger, M. Distribution of salt mixtures in a sandstone monument: Sources, transport and crystallization properties. In *European Commission Research Workshop: Origin, Mechanisms and Effects of Salts on Degradation of Monuments in Marine and Continental Environments. Protection and Conservation of the European Cultural Heritage. Research Report n° 4*; European Commission Research Workshop: Bari, Italy, 1996; pp. 241–246.

Disclaimer/Publisher's Note: The statements, opinions and data contained in all publications are solely those of the individual author(s) and contributor(s) and not of MDPI and/or the editor(s). MDPI and/or the editor(s) disclaim responsibility for any injury to people or property resulting from any ideas, methods, instructions or products referred to in the content.

Article

Archaeometric Study of the Mural Paintings by Saturnino Gatti and Workshop in the Church of San Panfilo, Tornimparte (AQ): The Study of Organic Materials in Original and Restored Areas

Alessia Andreotti [1,*], Francesca Caterina Izzo [2,*] and Ilaria Bonaduce [1]

[1] Dipartimento di Chimica e Chimica Industriale, Università di Pisa, Via Moruzzi 13, 56124 Pisa, Italy; ilaria.bonaduce@unipi.it
[2] Sciences and Technologies for the Conservation of Cultural Heritage, Department of Environmental Sciences, Informatics and Statistics, Ca' Foscari University of Venice, Via Torino 155/b, 30173 Venice, Italy
* Correspondence: alessia.andreotti@unipi.it (A.A.); fra.izzo@unive.it (F.C.I.)

Citation: Andreotti, A.; Izzo, F.C.; Bonaduce, I. Archaeometric Study of the Mural Paintings by Saturnino Gatti and Workshop in the Church of San Panfilo, Tornimparte (AQ): The Study of Organic Materials in Original and Restored Areas. *Appl. Sci.* **2023**, *13*, 7153. https://doi.org/10.3390/app13127153

Academic Editors: Rosina Celeste Ponterio, Mauro Francesco La Russa, Anna Galli, Maria Francesca Alberghina, Alessandro Re, Donata Magrini and Celestino Grifa

Received: 28 April 2023
Revised: 12 June 2023
Accepted: 13 June 2023
Published: 15 June 2023

Copyright: © 2023 by the authors. Licensee MDPI, Basel, Switzerland. This article is an open access article distributed under the terms and conditions of the Creative Commons Attribution (CC BY) license (https://creativecommons.org/licenses/by/4.0/).

Abstract: In the context of the archaeometrical study of Saturnino Gatti's wall paintings, a significant aspect concerned the study of the organic component to understand both the original binders used in the original areas and the products used for pictorial reintegration and restoration of the painted surfaces. Thanks to the results obtained from various non-invasive and multi-band imaging techniques, it was possible to define Gatti's original painting technique and identify the materials subsequently applied in significant samples. To this end, molecular analyses based on mass spectrometry were carried out. Different procedures in gas chromatography–mass spectrometry (GC-MS) and in pyrolysis coupled with gas chromatography–mass spectrometry (Py-GC-MS) were adopted. The analyses revealed a variety of organic materials on the mural paintings, most of which are from past restoration interventions and have synthetic origin. The overspread presence of paraffin is likely due to the application of a mineral wax-based coating/consolidant. In particular, the execution technique encompassed the use of tempera-based paints, while retouched areas were characterised by the presence of oil-based resins.

Keywords: proteins; lipids; paraffin; synthetic materials; paint binders

1. Introduction

This paper contributes to the Special Issue "Results of the II National Research project of AIAr: archaeometric study of the frescoes by Saturnino Gatti and workshop at the church of San Panfilo in Tornimparte (AQ, Italy)", in which the scientific results of the Second National Research Project conducted by members of the Italian Association of Archaeometry (AIAr) are discussed and collected. The cycle of frescoes is undoubtedly the masterpiece of the painter Saturnino Gatti. The wall paintings are framed by an architectural score, which creates illusionistic effects, running along the perimeter of the apse. A beautifully painted vault crowns the whole, and the sequence of images gives a beautiful 'cinemascope' effect.

The colourful scenes, subject to periodic restorations, are the subject of numerous studies, as highlighted by the vast literature.

Preliminary investigations of the painting technique, as well as the restoration materials of the pictorial cycle, were carried out using spectroscopic analysis and mainly non-invasive techniques, such as multiband imaging and single-spot techniques already optimized for works of art in cultural heritage [1–3].

The multispectral investigations carried out prior to the sampling campaign highlighted a diffuse presence of retouching, conservation and consolidation materials. Moreover, evident detachment and exfoliation phenomena were observed, together with a strong efflorescence. Chemical characterizations of the soluble salts by means of ion chromatography and attenuated total reflectance Fourier-transform infrared spectroscopy have been performed within the AIAr project [4]. Although the Renaissance pictorial cycle is of national interest, the pictorial technique has not yet been defined. Furthermore, the characterization of the consolidant and protective materials is necessary for the restoration, which aims to recover the legibility of the work, undermined by the degradation of the surface materials. For these purposes, within the research project, it was decided that twelve samples would be collected for chromatographic analyses. Currently, several works have demonstrated the capability of characterizing natural organic materials of different natures, such as lipid, protein, resinous and waxy materials with gas chromatography techniques coupled with mass spectrometry [5–10] and pyrolysis [11]. The pyrolysis analysis is also essential for the investigation of synthetic products used in restoration through the study of markers produced by the pyrolysis [12–14].

Despite the heavy and numerous restoration interventions, some areas were found to be of particular interest for the investigation of the paintings, such as residues of original gilding decorations, or original areas without overpainting or where aged coating films were less abundant.

2. Materials and Methods

2.1. Samples

The experimental sections are divided into two subsections: one dedicated at the identification of the original binding media and aged coatings, and the other to study the materials from the lacunae, as well as retouched and restored areas. Samples from the original paintings were collected mainly from the upper part of the vault and Area 1—Panel E (Figures 1a and 2), while for the consolidation/restoration ones they were collected mainly from Area 4—Panels A (Figure 1b), D and E (Figure 3). The bad state of conservation and severe legibility problems due to degraded pictorial varnishes, retouching and stuccos are clearly visible.

Figure 1. Visible image of Area 1—Panel E (**a**), and Area 4—Panel A (**b**).

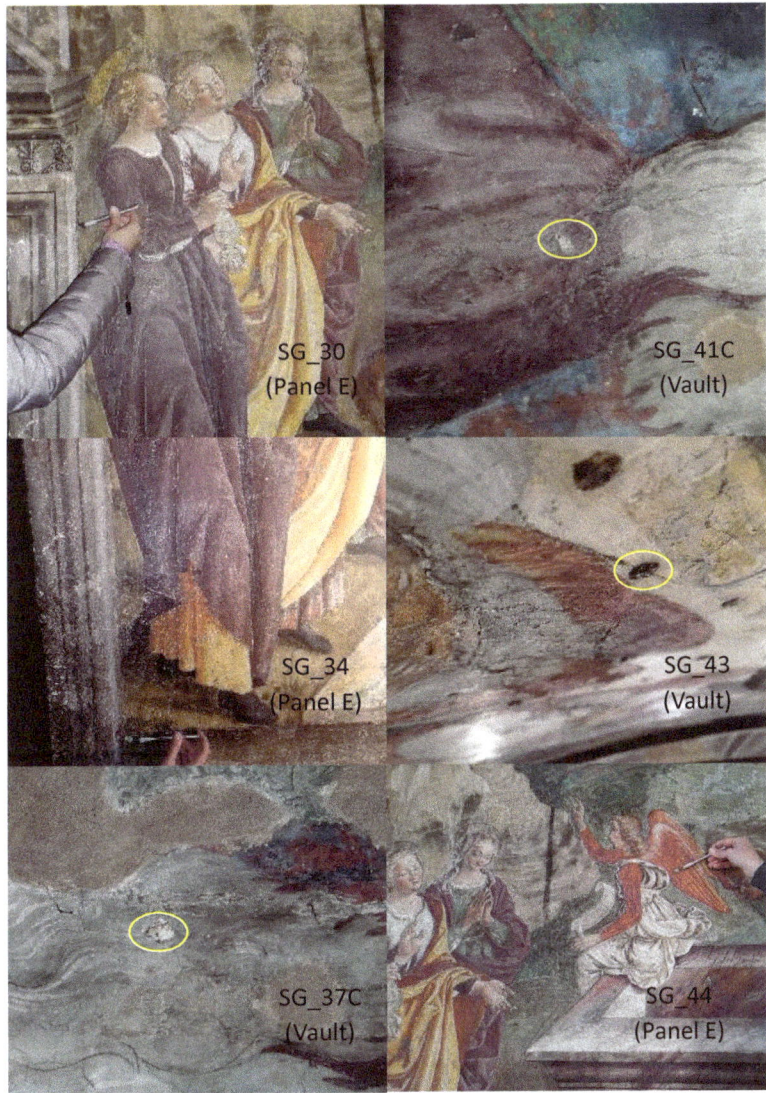

Figure 2. Sampling points of Area 1—Panel E and upper part of the vault for the characterisation of the original painting materials.

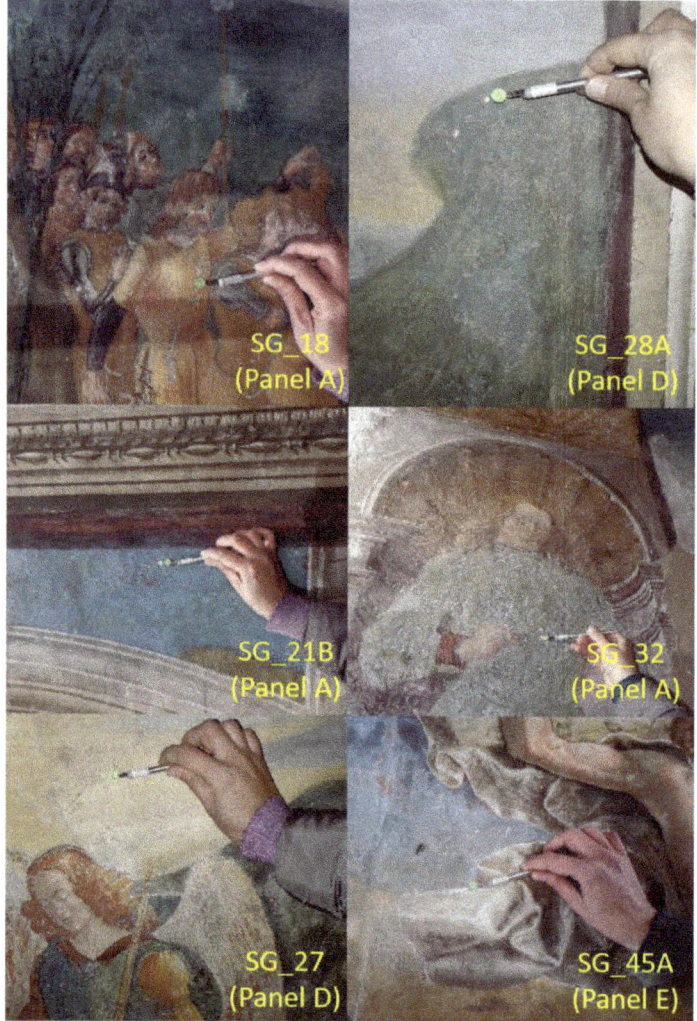

Figure 3. Sampling points of Area 4—Panels A, D and E for the characterisation of restoration materials.

2.1.1. Original Binding Media and Aged Coatings

Six samples were collected and analysed from the upper part of the vault and panel E (Figure 1a). Figure 2 shows the sampling points, and a brief sample description and the aim of studying the issues of the investigated areas are reported in Table 1. The areas, even if they were not as affected by thick overpainting and consolidants compared to the other investigated areas, still presented a film of degraded protective layers. Samples collected from these areas were analysed using gas chromatography–mass spectrometry (GC-MS), using an analytical procedure that enables the identification of natural organic materials based on lipids, proteins and terpenoids [15].

Table 1. Description of the samples collected for the identification of original binders and degraded superficial varnishes or coatings.

Sample	Area	Description	Aim of the Investigations
SG_30	Panel E	Fragments of white plaster + greyish pictorial finish taken from an area which under UV appears characterized by an uneven yellowish response. Macroscopically, the surface appears chromatically altered (with a spotted effect).	Identification of the degraded protective
SG_34	Panel E	Fragments of white plaster + grayish paint finish taken from an area that shows no fluorescence response under UV. Macroscopically, the surface appears glossy.	Identification of altered protective and/or other consolidating treatments undergone in the past
SG_37c	Vault, top part	Gray pictorial spread on a layer of degraded plaster, taken from a degraded area due to detachments and lifting of the surface layers.	Identification of organic binders and protective consolidants
SG_41C	Vault, top part	Purplish pictorial spread on a layer of degraded plaster, taken from a degraded area with lifting of the surface layers.	Identification of organic binders and protective consolidants
SG_43	Vault, top part	Relief layer of material of an organic nature selectively taken from the pictorial surface. These are tablets to simulate the golden decorations of the halos and robes of the Almighty (or relief base for the application of gold leaf), now detached.	Understanding of painting technique
SG_44	Panel E	Traces of a layer of preparation for the application of gold leaf for the creation of rays of the risen Christ. The surface of the layer taken is raised and tenaciously adheres to the plaster. Given the small number of traces still present, the sampling has been reduced to minimal quantities.	Understanding of the gold leaf application technique

2.1.2. Lacunae, Retouched and Restored Areas

Six samples were collected from Panels A, D and E, and the sampling points are shown in Figure 3. Table 2 reports a short description and visible images of the samples. Given the complexity of the pictorial fragments, which showed the co-presence of layers attributable to both original (or relatively ancient) work and subsequent restoration interventions, samples were divided into two aliquots. The first aliquot was analysed using GC-MS for the identification of lipid, terpenoid and waxy fractions, by means of an analytical procedure based on one-spot hydrolysis and transesterification [16,17]. The other aliquot was analysed with Py-GC-MS to verify the presence of proteinaceous and polysaccharide material [18] and detect synthetic polymers. In particular, analytical pyrolysis was carried out in combination with tetramethylammonium hydroxide (TMAH) to favour the detection of some organic binders by means of thermally assisted hydrolysis and methylation [19,20].

Table 2. Description and images of the samples collected for the identification of organic materials from lacunae, retouched and restored areas.

Sample	Area	Description	Image
SG_18A	Panel A	Yellow-orange pictorial layer and whitish preparation/primer, taken along an existing gap.	
SG_21B	Panel A	Blue pictorial layer on a red-brown layer (the so-called 'morellone') and underlying plaster fragments.	
SG_27A	Panel D	Yellow-blue pictorial application and white preparation/primer.	
SG_28A	Panel D	Dark green pictorial layer and fragments of plaster. The area also has pictorial integrations with glazes.	
SG_32	Panel A	Green pictorial layer and white preparation/primer, taken along the edge of a gap.	
SG_45A	Panel E	White layer ('lumeggiatura') (highlighting) applied on the underlying blue pictorial surface.	

2.2. Analytical Instrumentation

2.2.1. Instrumentation for the Analysis of Original Painting Materials

A 6890 gas chromatograph coupled with a 5975 single quadrupole mass-selective mass spectrometer (Agilent Technologies, Palo Alto, CA, USA) was used. The analytical procedure, the instrumental conditions and analytical parameters of which are reported elsewhere [15], allows the combined analysis of protein, lipid–resinous and wax content in samples. Samples were subjected to a multistep chemical pretreatment which involved several steps of extraction, hydrolysis, purification from inorganic species and silylation before GC-MS analysis.

2.2.2. Instrumentation for the Investigation of Restoration/Consolidation Treatment

- A Trace GC 1300 system equipped with an ISQ 7000 MS detector was used (ThermoFisher Scientific, Waltham, MI, USA) for the analysis of approximately 80–100 μg of sample. The transesterification reaction and the complete methodology are described elsewhere [16–18]. The data interpretation was performed using NIST and MS Search 1.7 libraries and ad hoc databases created by the authors. The Chromeleon 7 software was used for the data acquisition and processing.
- Thermally Assisted Hydrolysis and Methylation (THM)–Single Shot Pyrolysis–Gas Chromatography/Mass Spectrometry (TMH–SS-Py–GC/MS) was performed on small aliquots of samples (around 30–80 μg) in eco-cup pyrolysis crucibles. The samples were then treated with 3 μL of tetramethylammonium hydroxide (TMAH), 25%, in methanol.
- A PY-3030D pyrolizer (Frontier Lab, Koriyama, Japan), connected to a Trace 1310 gas chromatograph (ThermoFisher Scientific, Waltham, MA, USA) with an ISQ7000 mass spectrometer (ThermoFisher Scientific, Waltham, MA, USA), was used. The analytical parameters and software for collecting, processing and interpreting the mass spectral data are reported in the literature [19,20].

3. Results and Discussion

3.1. Original Binding Media, Aged Coatings and Consolidants

The presence of residues of synthetic origin (see following paragraph) caused an interference in the analysis of amino acids in sample SG_34; therefore, it was not possible to establish whether or not proteinaceous material was present. For all other samples, GC-MS analyses evidenced the presence of low amounts of proteinaceous material, which resulted above the limit of detection (LOD, 0.3 μg of the eleven amino acids quantified).

In sample SG_43, amino acids were present above the quantification limit (LOQ) of the analytical technique (0.6 μg in total). The relative amino acid content was thus subjected to multivariate statistical analysis according to the principal component method (PCA) together with a database of reference samples containing animal glue, egg and casein [15]. The resulting score plot is shown in Figure 4, which indicates that sample 43 contains egg. This result is of considerable importance because the sample had been taken from an area of the top part of the vault, selectively scraping the pictorial surface: we believe egg is an original binder used by the artist. The remaining samples present amino acids below the LOQ, and it is therefore not possible to identify the proteinaceous binder. Animal glue is absent, though, thus ruling out the presence of animal glue [21,22].

Beeswax is identified by the presence of relatively high amounts of palmitic acid and tetracosanoic acid and (ω-1) hydroxy even-numbered fatty acids (with the most abundant 15 hydroxyhexadecanoic acid) (Figure 5), long-chain linear alcohols with a number of even carbon atoms (Figure S1), α-(ω-1)diols and linear aliphatic hydrocarbons with an odd number of carbon atoms (Figure 6) [7,23]. On these bases, beeswax was clearly detected in all samples, with the exception of sample SG_44, and was particularly abundant in samples SG_37c and SG_43. Waxes were used as material for the consolidation and protection of wall paintings, as reported in many technical manuals. This is the case in Secco-Suardi's technical manual dated 1866, where restorers were advised to use mineral materials, such as paraffin, for the consolidation of wall paintings and painted facades [24–26]. The use of paraffin as a consolidating agent is evidenced by the restoration of the Pisa Cemetery in 1858 [23,27] and numerous restoration reports, including that of Venturini Papari for the Roman paintings in Pompeii [28].

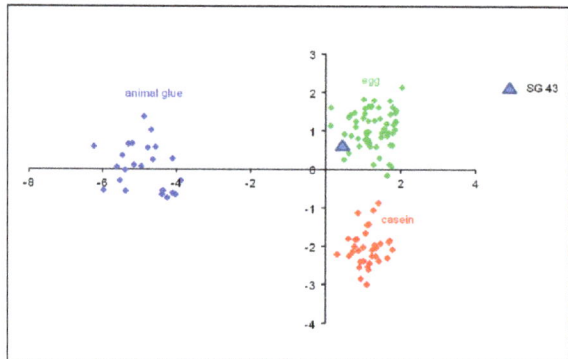

Figure 4. PCA score plot related to the percent amino acid profile of sample SG 43 (Vault) together with the database of profiles of reference samples containing animal glue, egg and casein.

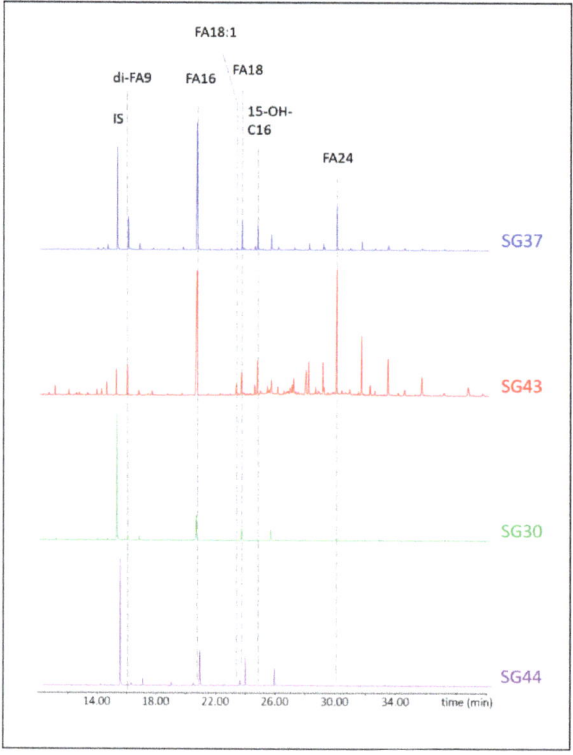

Figure 5. Extracted ion chromatogram of fragment ion m/z 129, characteristic of trimethylsilyl (TMS) esters. IS: internal standard: tridecanoic acid–TMS; di-FA9: azelaic acid–TMS; FA16: palmitic acid–TMS; FA18:1: oleic acid–TMS; FA18: stearic acid–TMS; 15-OHC-16: 15-hydroxy hexadecanoic acid–TMS; FA24: tetracosanoic acid–TMS.

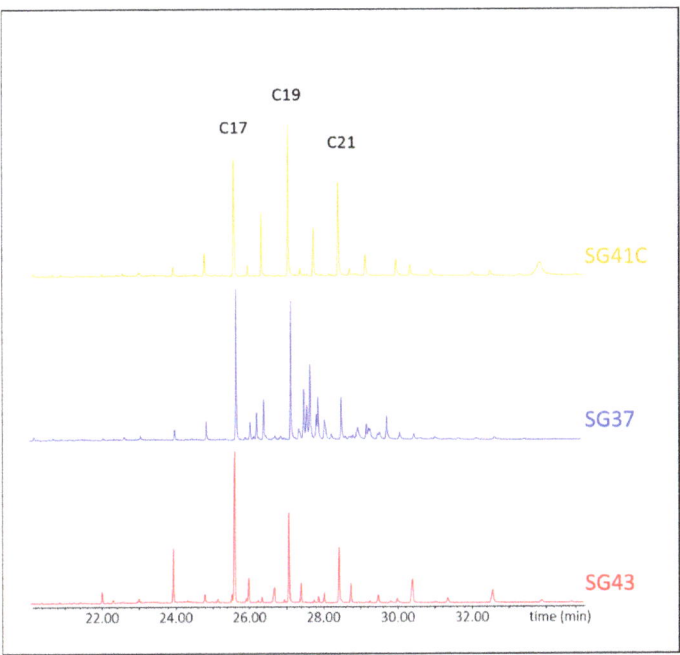

Figure 6. Extracted ion chromatogram of the fragment ion m/z 85, which is abundant in the mass spectra of linear aliphatic hydrocarbons. Cn indicates a linear aliphatic hydrocarbon with n carbon atoms.

Therefore, the use of paraffin and mineral waxes is also plausible in the case of Saturnino Gatti's wall paintings.

Beeswax makes the detection of glycerolipidic material difficult, as it substantially alters its chromatographic profile. The presence of relatively high amounts of azelaic acid, an oxidation product of polyunsaturated acids typical of glycerolipids of vegetable origin, points to the presence of an oxidized drying oil [29]. Traces were detected in both samples SG_30 and SG_44, but the chromatographic profile does not allow the determination of its nature, while relatively higher amounts were detected in samples SG_37c and SG_43 (SG_43 being a sample collected from the relief base for the application of gold leaf from the vault).

The analysis of the lipid–resinous fraction also evidenced the presence of a Pinaceae resin given the identification in the chromatogram of trimethylsilyl esters of dehydro-abietic acid and 7-oxo-dehydroabietic acid [21]. These are clearly visible in the chromatogram of the SG_43 sample (Figure S2), but detectable in the chromatograms of all the samples, except for sample SG_41C, due to an analytical interference which compromised the silylation reaction.

3.2. Lacunae, Retouched and Restored Areas

An extensive presence of paraffin was found in all the samples taken from Panels A, D and E, characterized by a series of linear alkanes with an number of carbon atoms higher than 25, and with a "Gaussian" shape [23,30] (Figure 7). Interestingly, paraffin was also found in samples taken from Panel E for the characterisation of the original binder, although it showed a different chromatographic profile and was particularly abundant in samples SG_30 and SG_44 (Figure 8). Paraffin was detected in all samples from Panel A, with the exception of sample SG_43.

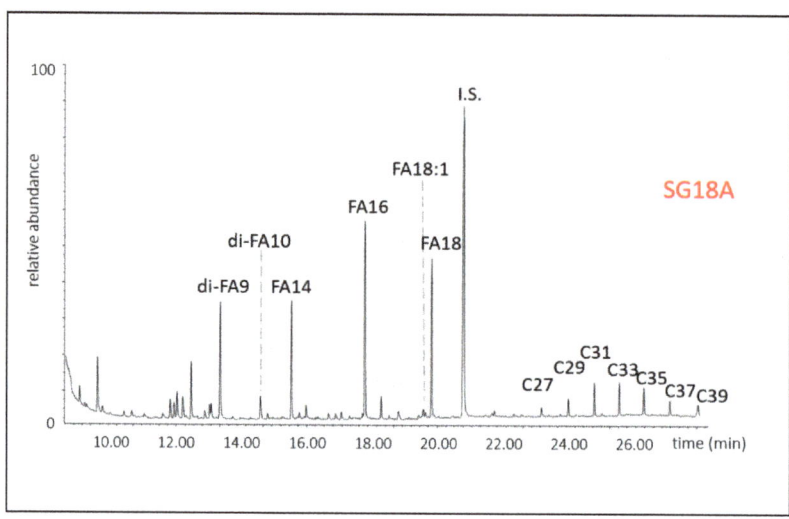

Figure 7. Total Ion Current (TIC) Chromatogram of sample SG_18A after methyl ester derivatization (ME). IS: internal standard: nonadecanoic acid–ME; di-FA9: azelaic acid–ME; di-FA10: sebacic acid–ME; FA14: myristic acid–ME; FA16: palmitic acid–ME; FA18:1: oleic acid–ME; FA18: stearic acid–ME; Cn linear aliphatic hydrocarbon with n atoms of carbon.

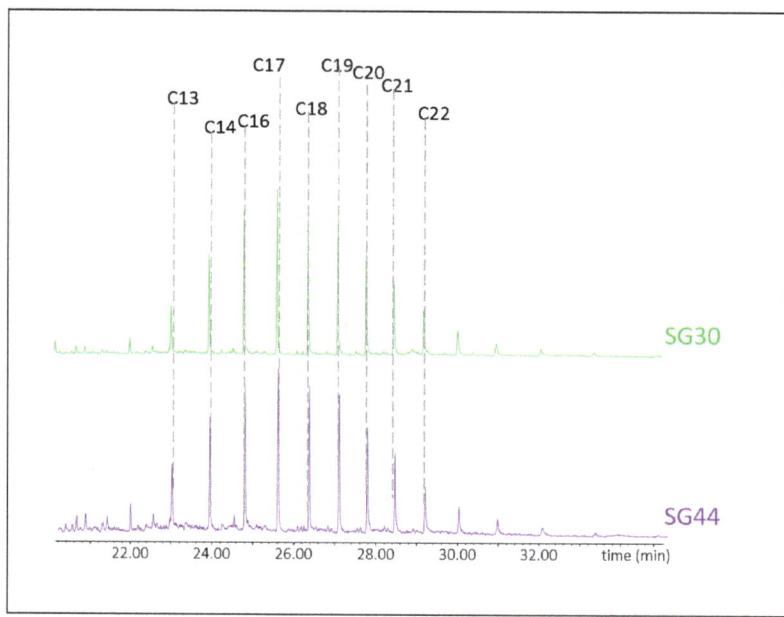

Figure 8. Chromatographic profiles of the ion extracted with m/z 85 for samples SG 44 and SG 30. Cn indicates a linear aliphatic hydrocarbon with n atoms of carbon.

A glycerolipid material was detected in samples SG_18A, SG_27A, SG_28A and SG_32, based on the identification of saturated monocarboxylic fatty acids (palmitic and stearic) and dicarboxylic acids (sebacic, azelaic and suberic acids) (Figure 6). The chromatographic profile, and in particular the relatively high content of dicarboxylic acids, indicates the presence of a siccative oil [29]. Moreover, SG_18A and SG_32 still present oleic acid, a monounsaturated fatty acid generally still present in not completely mature oil paint films. Its presence suggests that the drying oil found in the samples is most likely a later material rather than an original one used by Saturnino Gatti.

Myristic acid, a saturated fatty acid naturally present in many vegetable lipid materials, is ubiquitous. Its relative abundance in the chromatographic profile strongly suggests that it is not derived from the drying oil, but it could be present as an additive (such as a surfactant or emulsifier).

A plant resin of the Pinaceae family was identified in samples SG_27 (both yellow and orange) and SG_32, based on the detection of abietic acid and its oxidation products [21] (dehydroabietic acid, 15-oxo-dehydroabietic acid and 7-oxo-dehydro-abietic acid). This terpenoid material, not traditionally used as a binder in wall paintings, could be present as an additive of the binder used for retouching or in oil-resinous coatings. The presence of abietic acid also emphasises that the painting is relatively fresh and therefore not of the same period as the creation of the painting cycle.

Synthetic polymers were also found. In particular, an oil-modified alkyd-based resin was identified in samples SG_28A and SG_45A, based on the detection of orthophthalic acid, benzoic acid, glycerol and a drying oil [12]. An acrylic-styrene resin was identified in sample SG_21B, thanks to the main presence of styrene, butyl acrylate, n-butyl methacrylate and α-methylstyrene [31,32].

These synthetic materials are likely present as binders of the paints used for the reintegration, being amongst the most used contemporary paints.

For the identification of possible proteinaceous and polysaccharidic fractions remaining from the original binder, the samples were also subjected to analysis with Py-GC-MS in Double Shot mode (350–650 °C) as described in the literature [33]. The analyses were not conclusive, however, as the typical markers of the proteins present in casein or egg-based paints were not detected. This aspect could be linked to natural aging and cross-linking processes, as well as to the interference due to the inorganic component of the samples and the many materials used in restoration and painting reintegration.

In conclusion, it can be stated that the samples under investigation appear to have been taken from areas that were significantly, if not completely, repainted. Therefore, it was not possible to identify the original organic binder, while the composition of the different paints used for reintegration was clarified.

4. Conclusions

The analysis of organic materials revealed the presence of a variety of natural and synthetic materials, indicating numerous interventions during the years.

It is difficult to establish the binding media of the original painting layers, due to extensive contaminations and co/presence of many sources of organic materials, both of natural and synthetic origin. Despite this, the presence of proteinaceous material, most likely egg, was observed in the samples coming from the vault, suggesting the use by Saturnino Gatti of an egg tempera, at least in some areas.

A siccative oil was detected and clearly identified in several samples collected near some lacunae together with synthetic resins, suggesting that this is a restoration material. Additionally, terpenic materials and alkyd resins were detected near/close to lacunae and retouched areas, and they could be associated with the use of the so-called «colori da ritocco» [27], which are paints specifically designed to be used for retouching paintings in restoration interventions.

Beeswax and more than one type of paraffin wax were extensively detected in the sections of the wall paintings analysed.

In conclusion, the results obtained from the organic analyses provided a great deal of information regarding the materials presumably used in the creation of the pictorial cycle, as well as in the various phases of reworking and restoration that took place over the years (testifying also the importance of the pictorial cycle and the desire to preserve it over time).

Together with the results obtained from the various analytical techniques presented in this Special Issue, the knowledge relating to this unique pictorial cycle has been expanded, and will certainly be put to the best use for its conservation and valorisation.

Supplementary Materials: The following supporting information can be downloaded at: https://www.mdpi.com/article/10.3390/app13127153/s1.

Author Contributions: Formal analysis, A.A. and F.C.I.; Investigation, A.A. and F.C.I.; Writing—original draft, A.A., F.C.I. and I.B.; Visualization, A.A., F.C.I. and I.B.; Writing—revision, A.A., F.C.I. and I.B. All authors have read and agreed to the published version of the manuscript.

Funding: This work was performed in the framework of the Project Tornimparte—"Archeometric investigation of the pictorial cycle of Saturnino Gatti in Tornimparte (AQ, Italy)" sponsored in 2021 by the Italian Association of Archeometry (AIAr).

Institutional Review Board Statement: Not applicable.

Informed Consent Statement: Not applicable.

Data Availability Statement: Data are available upon reasonable request.

Acknowledgments: Francesca Caterina Izzo would like to thank the Patto per lo Sviluppo della Città di Venezia (Comune di Venezia) for the support in the research.

Conflicts of Interest: The authors declare no conflict of interest.

References

1. Magrini, D.; Bracci, S.; Cantisani, E.; Conti, C.; Rava, A.; Sansonetti, A.; Shank, W.; Colombini, M. A multi-analytical approach for the characterization of wall painting materials on contemporary buildings. *Spectrochim. Acta-Part A Mol. Biomol. Spectrosc.* **2017**, *173*, 39–45. [CrossRef] [PubMed]
2. Bracci, S.; Iannaccone, R.; Magrini, D. The application of multiband imaging integrated with non-invasive spot analysis for the examination of archeological stone artifacts. *CONSERVATION 360°* **2020**, 141–160. Available online: https://monografias.editorial.upv.es/index.php/con_360/article/view/71 (accessed on 12 February 2022).
3. Bracci, S.; Cantisani, E.; Conti, C.; Magrini, D.; Vettori, S.; Tomassini, P.; Marano, M. Enriching the knowledge of Ostia Antica painted fragments: A multi-methodological approach. *Spectrochim. Acta Part A Mol. Biomol. Spectrosc.* **2022**, *265*, 120260. [CrossRef] [PubMed]
4. Comite, V.; Bergomi, A.; Lombardi, C.A.; Borelli, M.; Fermo, P. Characterization of Soluble Salts on the Frescoes by Saturnino Gatti in the Church of San Panfilo in Villagrande di Tornimparte (L'Aquila). *Appl. Sci.* **2023**, *13*, 6623. [CrossRef]
5. Da Filicaia, E.G.; Evershed, R.P.; Peggie, D.A. Review of recent advances on the use of mass spectrometry techniques for the study of organic materials in painted artworks. *Anal. Chim. Acta* **2023**, *1246*, 340575. [CrossRef] [PubMed]
6. Calvano, C.D.; van der Werf, I.D.; Palmisano, F.; Sabbatini, L. Revealing the composition of organic materials in polychrome works of art: The role of mass spectrometry-based techniques. *Anal. Bioanal. Chem.* **2016**, *408*, 6957–6981. [CrossRef]
7. Mazurek, J.; Svoboda, M.; Schilling, M. GC/MS characterization of beeswax, protein, gum, resin, and oil in romano-egyptian paintings. *Heritage* **2019**, *2*, 1960–1985. [CrossRef]
8. Castellá, F.; Pérez-Estebanez, M.; Mazurek, J.; Monkes, P.; Learner, T.; Niello, J.F.; Tascon, M.; Marte, F. A multi-analytical approach for the characterization of modern white paints used for Argentine concrete art paintings during 1940–1960. *Talanta* **2020**, *208*, 120472. [CrossRef]
9. Rogge, C.E.; Mazurek, J.; Schilling, M. The Nucleus of Color: Analysis of Hélio Oiticica's Studio Materials. *Stud. Conserv.* **2022**. [CrossRef]
10. Schilling, M.R. Paint media analysis. In *Scientific Examination of Art: Modern Techniques in Conservation and Analysis*; National Academies Press: Washington, DC, USA, 2005; pp. 186–205. [CrossRef]
11. Chiavari, G.; Galletti, G.C.; Lanterna, G.; Mazzeo, R. The potential of pyrolysis—Gas chromatography/mass spectrometry in the recognition of ancient painting media. *J. Anal. Appl. Pyrolysis* **1993**, *24*, 227–242. [CrossRef]
12. Wei, S.; Pintus, V.; Schreiner, M. A comparison study of alkyd resin used in art works by Py-GC/MS and GC/MS: The influence of aging. *J. Anal. Appl. Pyrolysis* **1970**, *104*, 441–447. [CrossRef]
13. Wei, S.; Pintus, V.; Schreiner, M. Photochemical degradation study of polyvinyl acetate paints used in artworks by Py-GC/MS. *J. Anal. Appl. Pyrolysis* **2012**, *97*, 158–163. [CrossRef] [PubMed]
14. Peris-Vicente, J.; Baumer, U.; Stege, H.; Lutzenberger, K.; Gimeno Adelantado, J.V. Characterization of Commercial Synthetic Resins by Pyrolysis-Gas Chromatography/Mass Spectrometry: Application to Modern Art and Conservation. *Anal. Chem.* **2009**, *81*, 3180–3187. [CrossRef] [PubMed]

15. Bonaduce, I.; Cito, M.; Colombini, M.P. The development of a gas chromatographic-mass spectrometric analytical procedure for the determination of lipids, proteins and resins in the same paint micro-sample avoiding interferences from inorganic media. *J. Chromatogr. A* **2009**, *1216*, 5931–5939. [CrossRef] [PubMed]
16. Izzo, F.C.; Lodi, G.C.; Vázquez de Ágredos Pascual, M.L. New insights into the composition of historical remedies and pharmaceutical formulations: The identification of natural resins and balsams by gas chromatographic-mass spectrometric investigations. *Archaeol. Anthropol. Sci.* **2021**, *13*, 2. [CrossRef]
17. Izzo, F.C.; Ferriani, B.; Van den Berg, K.J.; Van Keulen, H.; Zendri, E. 20th century artists' oil paints: The case of the Olii by Lucio Fontana. *J. Cult. Herit.* **2014**, *15*, 557–563. [CrossRef]
18. Morales Toledo, E.G.; Raicu, T.; Falchi, L.; Barisoni, E.; Piccolo, M.; Izzo, F.C. Critical Analysis of the Materials Used by the Venetian Artist Guido Cadorin (1892–1976) during the Mid-20th Century, Using a Multi-Analytical Approach. *Heritage* **2023**, *6*, 600–627. [CrossRef]
19. Izzo, F.; Van Keulen, H.; Carrieri, A. Assessing the Condition of Complex Poly-Material Artworks by Py-GC-MS: The Study of Cellulose Acetate-Based Animation Cels. *Separations* **2022**, *9*, 131. [CrossRef]
20. Van Keulen, H.; Schilling, M. AMDIS & EXCEL: A Powerful Combination for Evaluating THM-Py-GC/MS Results from European Lacquers. *Stud. Conserv.* **2019**, *64*, S74–S80.
21. Colombini, M.P.; Andreotti, A.; Bonaduce, I.; Modugno, F.; Ribechini, E. Analytical strategies for characterizing organic paint media using gas chromatography/mass spectrometry. *Acc. Chem. Res.* **2010**, *43*, 715–727. [CrossRef]
22. Dallongeville, S.; Garnier, N.; Rolando, C.; Tokarski, C. Proteins in art, archaeology, and paleontology: From detection to identification. *Chem. Rev.* **2016**, *116*, 2–79. [CrossRef]
23. Bonaduce, I.; Colombini, M.P. Characterisation of beeswax in works of art by gas chromatography-mass spectrometry and pyrolysis-gas chromatography-mass spectrometry procedures. *J. Chromatogr. A* **2004**, *1028*, 297–306. [CrossRef]
24. Secco-Suardo, G. *Manuale Ragionato per la Parte Meccanica Dell'arte del Restauratore dei Dipinti*; Forgotten Books: London, UK, 1866.
25. Mora, P.; Mora, L.; Philippot, P. *Conservazione delle Pitture Murali*, 2nd ed.; HOEPLI: Milan, Italy, 2003.
26. Bertorello, C. *Materiali a Confronto sui Dipinti Murali Nell'esperienza dei Restauratori Tra'800 e'900*; Bollettino d'arte: Rome, Italy, 1996.
27. Vanni, T. *Le Decorazioni Pittoriche del Centro di Portogruaro*; Associazione Accordi Portogruaro: Portogruaro, Italy, 2004.
28. Prisco, G. Tecnica esecutiva e conservazione delle pitture murali di epoca romana. Il dibattito tra fine '800 e prima metà del '900. *Boll. ICR* **2013**, *27*, 50–69.
29. La Nasa, J.; Modugno, F.; Degano, I. Liquid chromatography and mass spectrometry for the analysis of acylglycerols in art and archeology. *Mass Spectrom. Rev.* **2021**, *40*, 381–407. [CrossRef] [PubMed]
30. Asperger, A.; Engewald, W.; Fabian, G. Advances in the analysis of natural waxes provided by thermally assisted hydrolysis and methylation (THM) in combination with GC/MS. *J. Anal. Appl. Pyrolysis* **1999**, *52*, 51–63. [CrossRef]
31. Pintus, V.; Schreiner, M. Characterization and identification of acrylic binding media: Influence of UV light on the ageing process. *Anal. Bioanal. Chem.* **2011**, *399*, 2961–2976. [CrossRef]
32. Fardi, T.; Pintus, V.; Kampasakali, E.; Pavlidou, E.; Schreiner, M.; Kyriacou, G. Analytical characterization of artist's paint systems based on emulsion polymers and synthetic organic pigments. *J. Anal. Appl. Pyrolysis* **2018**, *135*, 231–241. [CrossRef]
33. Orsini, S.; Parlanti, F.; Bonaduce, I. Analytical pyrolysis of proteins in samples from artistic and archaeological objects. *J. Anal. Appl. Pyrolysis* **2017**, *124*, 643–657. [CrossRef]

Disclaimer/Publisher's Note: The statements, opinions and data contained in all publications are solely those of the individual author(s) and contributor(s) and not of MDPI and/or the editor(s). MDPI and/or the editor(s) disclaim responsibility for any injury to people or property resulting from any ideas, methods, instructions or products referred to in the content.

Article

Indoor Microclimate Analysis of the San Panfilo Church in Tornimparte, Italy

Silvia Ferrarese [1,*], Davide Bertoni [1], Alessio Golzio [1], Luca Lanteri [2], Claudia Pelosi [2] and Alessandro Re [1,3]

[1] Department of Physics, University of Turin, 10125 Turin, Italy; davide.bertoni@unito.it (D.B.); alessio.golzio@unito.it (A.G.); alessandro.re@unito.it (A.R.)
[2] Department of Economics, Engineering, Society and Business Organization, University of Tuscia, 01100 Viterbo, Italy; llanteri@unitus.it (L.L.); pelosi@unitus.it (C.P.)
[3] Istituto Nazionale di Fisica Nucleare (INFN), Turin Division, 10125 Turin, Italy
* Correspondence: silvia.ferrarese@unito.it

Abstract: This work presents the results of a microclimatic analysis in the church of San Panfilo in Tornimparte, in the Abruzzo region, Italy. The church of San Panfilo, dating from the XII–XIII century, preserves in the presbytery some marvelous frescoes of Saturnino Gatti (1494). A measurement campaign was organized from February 2021 to April 2022 with the aim of investigating the microclimate inside the church in view of the organization of fresco restoration. The monitoring activity was performed with intensive measurements on specific days during the year and with continuous measurements throughout the whole year. The main microclimatic parameters, relative humidity and temperature, were monitored in various sites inside and outside the church. Some physical quantities, such as dew point temperature, dew point spread and specific humidity, were computed from measured data. Measured and computed data permitted to evaluate the daily and monthly values, their evolution during the year, the daily fluctuations and delay times caused by the building. The resulting discussion allowed to identify potentially dangerous events for the conservation of the frescoes. Moreover, the historic climate inside the church was detected.

Keywords: microclimate; cultural heritage conservation; historic church; indoor environmental monitoring

1. Introduction

Some historic buildings have miraculously been preserved to the present day, overcoming many adversities such as wars, fires and earthquakes. In addition, some historic buildings preserve wonderful artworks that amaze, comfort and delight people who can admire them. The conservation of artistic and cultural heritage is therefore a duty; moreover, due to the proximity of beauty, it is also an enjoyable commitment for professionals.

The aim of studying microclimate for cultural heritage is to investigate the environmental conditions around a work of art and to individuate critical and dangerous conditions [1,2]. In fact, works of art can be conserved over time if the surrounding environment does not favor the processes of degradation. Microclimatic monitoring is therefore essential to know the environmental situation and possibly to propose solutions or improvements in order to promote conservation.

In order to establish safe ranges in microclimatic variability, some standards have been proposed, such as UNI 10829 [3] and EN 15757 [4]. The UNI 10829 standard established the methodology to evaluate the microclimate and suggested the range of variability for microclimatic parameters in order to properly preserve artworks. This norm has the advantage of giving practical guidance in organizing a measurement campaign and in fixing the safe zone using threshold values for 33 classes of artworks; its disadvantage lies in the rigid safe ranges that are not applicable in every environment. The EN 15757 standard overcame this difficulty with the definition of the historic climate that considers the environment conditions where the works of art are acclimatized. In order to compare different microclimatic conditions, several indexes have been defined (such as [5–7]); they are easily computed using more than one microclimatic parameter and provide comprehensive information about different sites. Despite the availability of these useful instruments, the debate around safe zones is still open.

Historical buildings and churches have conserved ancient artworks for centuries, and they are themselves cultural heritage. Their indoor microclimates are favorable to conservation mainly as the thermal inertia of their massive thick walls mitigates external fluctuations. In the last 30 years, many historic buildings have been monitored and studied (among others, [8–13]) with the aim to investigate the microclimatic conditions and to suggest improvements to restorers. Recently, the microclimate inside small and ancient churches preserving priceless heritage [14–16] has been analyzed.

In this work, we present the monitoring campaign performed in the church of San Panfilo in Tornimparte (L'Aquila, Italy) with the aim of detecting the microclimatic situation and identifying any critical conditions for conservation.

The parish church of Tornimparte dates back to the second part of the XII century or the beginning of the XIII century, and during its long history, it underwent profound renovations as a consequence of earthquakes and fires. The church now presents four naves with some altars and a painted deep apse.

Saturnino Gatti depicted the cycle of frescoes in the apse from 1491 to 1494 [17,18]. The cycle includes a representation of the glory of God in the vault and scenes of the Christ's passion and resurrection on the side walls. In the central part of the apse, it is likely that the crucifixion was represented, but the scene was lost in 1922 with the opening of a window [18].

In lateral naves, some frescoes, paintings and statues are preserved; moreover, outside the church, part of the external portico is frescoed with medieval paintings on the external church walls and with a Madonna and Child in the lunette above the main door.

In this work, the monitoring activity is described, and the results are discussed. Monitored physical quantities were temperature and relative humidity inside and outside the church. The monitoring period lasted 15 months in order to identify hourly, daily and seasonal variability in measured data.

This paper contributes to the Special Issue "Results of the II National Research project of AIAr: archaeo-metric study of the frescoes by Saturnino Gatti and workshop at the church of San Panfilo in Tornimparte (L'Aquila, Italy)" in which the scientific results of II National Research Project conducted by members of the Italian Association of Archaeometry (AIAr) are discussed and collected. For in-depth details on the aims of the project, see the Introduction of the Special Issue [19].

2. Materials and Methods

2.1. San Panfilo Church in Tornimparte

San Panfilo church (latitude: 42.28864° N, longitude: 13.30136° E) is located in the municipality of Tornimparte, province of L'Aquila, Abruzzo region, in central Italy (Figure 1). San Panfilo church is an isolated building arranged along the south–north direction with the main facade facing north and the apse facing south (Figure 2). The church faces a large churchyard, is enclosed by a wall, and the facade is preceded by a portico. The building is not symmetric; in fact, it presents four naves with a main central nave aligned from the main door to the presbytery, a lateral nave on the left and two lateral naves on the right.

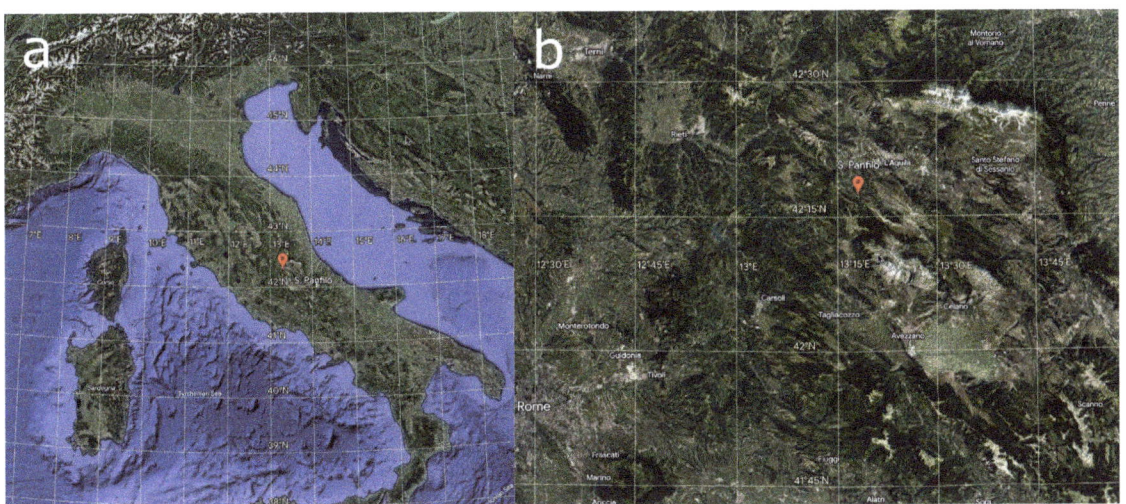

Figure 1. (**a**) Localization of San Panfilo church in Tornimparte in central Italy and (**b**) in Abruzzo region.

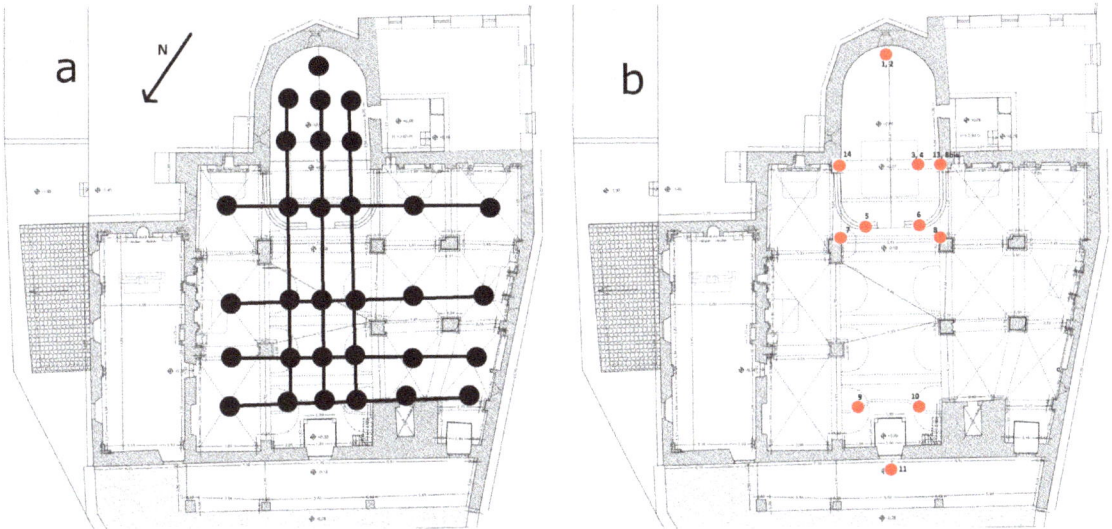

Figure 2. Church plan: (**a**) sites on spatial grid for intensive measurements (black points and lines); (**b**) positions of instruments for continuous monitoring (red points).

A wooden balustrade borders the presbytery, which is furnished with a stone altar, the seat of the celebrant and the concelebrants, a wooden crucifix on a wrought iron base and a glass showcase near the central window containing a silver processional cross from the 1600 s. The choir is located inside the church, above the main door.

Light through window glasses illuminates the church: a series of windows are in the upper walls of main nave, three windows in the lateral walls and two windows in the apse on the left side and in the center. An artificial lighting system was recently installed; it consists of a number of LED lights over the column capitals. In the church, the air exchange is guaranteed through the main door and through the opening of two lateral windows in the external walls.

The church is equipped with a heating system; nevertheless, it was turned off during the whole monitoring activity. Parishioners frequent San Panfilo: the church is used for Sunday liturgy and for services related to funerals and weddings.

2.2. Monitoring Campaign

Monitoring activity started in February 2021 and ended in April 2022. The two main microclimatic parameters, temperature and relative humidity, were measured inside and outside the church. The monitoring activity was organized into two main phases: intensive measurements and continuous measurements.

Intensive measurements were performed on specific days (18 February 2021, 6 July 2021, 22 November 2021, 22 April 2022) chosen along the year with the aim to inspect in detail the situation inside the church through the seasons. Temperature and relative humidity were measured two times a day (in the morning and early afternoon) at two vertical levels (1.2 m and 2.0 m from the floor) on a regular grid (Figure 2a).

On 22 April 2022, some additional measures of temperature and relative humidity were performed at four levels from the floor (0.5, 1.2, 2.0, 3.0 m) in the apse. Moreover, on 22 November 2021 and 22 April 2022, the apse wall temperature was measured with an infrared thermometer.

Continuous measurements were performed from 20 February 2021 to 18 April 2022 in several sites in the church and two sites outside the church (Figure 2b, Table 1) at a monitoring frequency of one datum every 10, 15 or 60 min. The apse was monitored using two sensors placed on the glass showcase (positions 1 and 2), two sensors tied to the crucifix (positions 3 and 4) and three sensors placed near the walls on two little shelves (positions 8bis, 13, and 14). In the main nave the monitoring sites were at the balustrade (positions 5 and 6), on the pulpit (position 7), on the first column capital (position 8) and on the choir (positions 9 and 10). The outdoor environment was monitored using a sensor below the external portico (position 11) and near the meteorological station located at Colle San Vito (latitude: 42.290639° N, longitude: 13.289556° E) at a distance of about 1 km from San Panfilo church. This station is managed by the Centro Funzionale and Ufficio Idrografico of Regione Abruzzo.

Table 1. Positions of sensors during continuous monitoring.

Positions (Figure 2)	Site	Height from The Floor (cm)	Acquisition Time Intervals (Minutes)	Acquisition Time Period	Instrument
1	case (left)	86	10	20 February 2021 to 18 April 2022	HOBO-UX100-011
2	case (right)	86	10	20 February 2021 to 18 April 2022	HOBO-UX100-011
3	crucifix (bottom)	110	10	20 February 2021 to 18 April 2022	HOBO-UX100-011
4	crucifix (top)	207	10	20 February 2021 to 7 April 2022	HOBO-UX100-011
5	balustrade (left)	56	10	20 February 2021 to 18 April 2022	HOBO-UX100-011
6	balustrade (right)	56	10	20 February 2021 to 18 April 2022	HOBO-UX100-011
7	pulpit	214	10	20 February 2021 to 18 April 2022	HOBO-UX100-011
8	column	240	10	20 February 2021 to 22 November 2021	HOBO-UX100-011
8bis	apse (right)	328	10	22 November 2021 to 18 April 2022	HOBO-UX100-011
9	choir (left)	278	10	20 February 2021 to 18 April 2022	HOBO-UX100-011
10	choir (right)	278	10	6 July 2021 to 18 April 2022	HOBO-UX100-011
11	portico	268	10	20 February 2021 to 30 March 2022	HOBO-UX100-011
12	colle San Vito meteorological station	/	15	20 February 2021 to 18 April 2022	/
13	apse (right)	328	60	1 February 2021 to 31 January 2022	Testo 175-H2
14	apse (left)	315	60	1 February 2021 to 31 January 2022	Testo 177-H1

Some microclimatic parameters such as specific humidity, dew point temperature and dew point spread were computed from temperature and relative humidity data.

2.3. Instrumentation

The intensive measurements of temperature and relative humidity were made using a portable thermo-hygrometer (manufactured by Testo Spa, model 625, Settimo Milanese, Milan, Italy, Figure 3a), whereas the wall temperature was measured with a infrared thermometer (manufactured by Fluke, model 62 Mini, Everett, WA, USA, Figure 3b). The continuous measurements were collected using eleven thermo-hygrometers (manufactured by Onset, model HOBO-UX100-011, Bourne, MA, USA, Figure 3c) working at the frequency of one datum every 10 min and two thermo-hygrometers (manufactured by Testo Spa, models 177-H1 and 175-H2, Settimo Milanese, Milan, Italy, Figure 3d) at the frequency of 1 datum every 60 min. The temperature and relative humidity accuracy, resolution and range of the used sensors are summarized in Table 2.

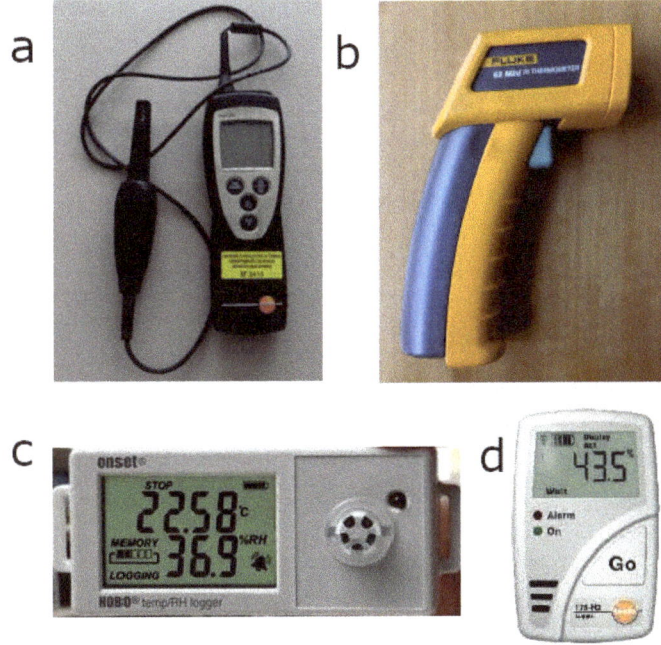

Figure 3. Instruments: (**a**) Testo 625 thermo-hygrometer, (**b**) Fluke 62 Mini infrared thermometer, (**c**) HOBO-UX100-011 thermo-hygrometer, and (**d**) Testo 175-H2 thermo-hygrometer.

Table 2. Instruments description.

Instrument	Temperature Accuracy (°C)	Temperature Resolution (°C)	Temperature Range (°C)	Relative Humidity Accuracy (%)	Relative Humidity Resolution (%)	Relative Humidity Range (%)
HOBO-UX100-011	0.21	0.024	−20 to 70	2.5	0.05	1 to 95
Testo 625	0.5	0.1	−10 to 60	2.5	0.1	0 to 100
Fluke 62 Mini	0.5	0.2	−30 to 500	-	-	-
Testo 177-H1/175-H2	0.5	0.1	−20 to 70	3.0	0.1	0 to 100

2.4. Computation of Specific Humidity, Dew Point Temperature, and Dew Point Spread

Specific humidity is defined as the ratio between the mass of water vapor and the mass of moist air. In microclimatic analysis this physical quantity is useful to recognize the presence of evaporation, condensation, the path of an air mass, leakage, window openings and the presence of people [1]. Specific humidity (SH) was computed using the following formula:

$$SH = 0.62197 \cdot \frac{e}{p-e} 100$$

where e is the vapor pressure and p is the pressure at the altitude of Tornimparte. The vapor pressure e was computed using the equation:

$$e = \frac{RH \cdot e_w}{100}$$

with RH representing relative humidity and e_w representing the saturation vapor from Bolton parametrization [20]:

$$e_w = 6.112 \cdot \exp\left(\frac{17.67 \cdot T}{T + 243.5}\right)$$

where T is the measured temperature (°C).

The dew point temperature is defined as the temperature to which a parcel of moist air must be cooled at constant atmospheric pressure and constant water vapor content in order for saturation to occur. Dew point temperature (T_d) was computed using the relation:

$$T_d = b \frac{a \cdot T + (b+T) \cdot \log\left(\frac{RH}{100}\right)}{a \cdot b - (b+T) \cdot \log\left(\frac{RH}{100}\right)}$$

where a = 7.5 and b = 237.3 °C are the Magnus and Tetens coefficients for vapor in equilibrium with the liquid phase [1].

Dew point spread is computed as the difference between the temperature and the dew point temperature. It shows how close or far the air temperature is from the dew point temperature. In conservation science, the dew point spread gives information about the possibility of condensation over the surfaces.

3. Results

3.1. Intensive Monitoring

The analysis of intensive monitoring data was carried out considering the horizontal fields of micrometeorological parameters at 1.2 m and 2.0 m from the floor (Section 3.1.1) and the vertical profiles measured in the apse (Section 3.1.2).

3.1.1. Analysis on a Horizontal Spatial Grid

Measured data on temperature and relative humidity and computed values for specific humidity and dew point spread were used to obtain the horizontal fields. A total of 52 horizontal maps were examined. The results on 6 July 2021 and 22 November are representative, respectively, of the summer and winter seasons and are shown in Figures 4 and 5.

Results show that values of temperature, specific humidity and dew point spread inside the church are homogeneous and depend on seasonality. Relative humidity values belong to the same range (about 60–67%) in every season but not in summer, when the relative humidity values are lower than 55% (Figures 4b and 5b).

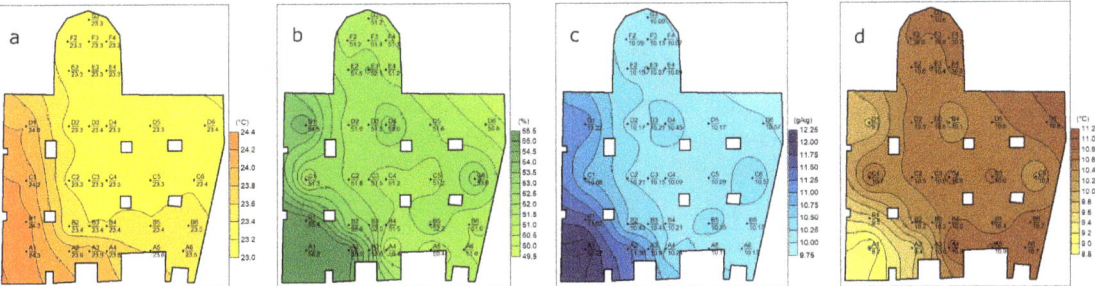

Figure 4. (a) Temperature, (b) relative humidity, (c) specific humidity and (d) dew point spread at 2.0 m from the floor on 6 July 2021 at 9:00 UTC+1.

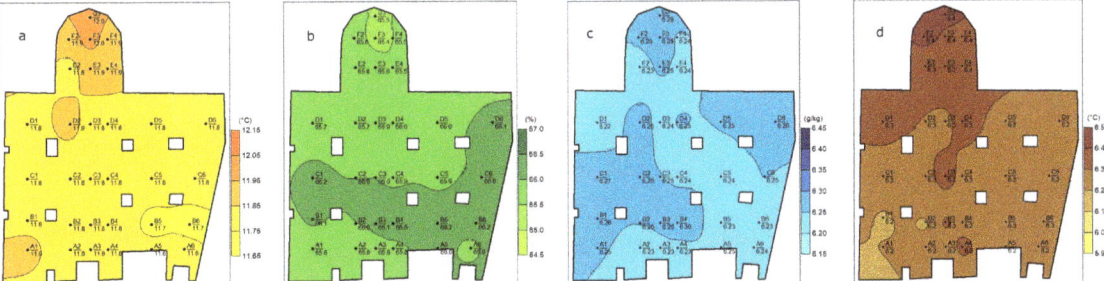

Figure 5. (a) Temperature, (b) relative humidity, (c) specific humidity and (d) dew point spread at 1.2 m from the floor on 22 November 2021 at 14:30 UTC+1.

In February, November and April, the horizontal variability for temperature is lower than 0.5 °C; for relative humidity, it is lower than 2.5%; for specific humidity, it is lower than 0.25 g/kg; for dew point spread, it is lower than 0.5 °C. Therefore, the horizontal maps show values depending on seasonality, but the amplitude of scale is the same. In July, the microclimatic horizontal variability is higher (1.1 °C for temperature, 7.8% for relative humidity, 2.2 g/kg for specific humidity and 2.1 °C for dew point distance) as a consequence of some values near the main church door on the left measured in the morning (Figure 4). In this area, a review of instability has reported the presence of moisture infiltrations.

In the presbytery, the horizontal variability in microclimatic parameters is very low (0.3 °C for temperature, 1.8% for relative humidity, 0.4 g/kg for sensible humidity and 0.6 °C for dew point distance), close to instrument accuracy.

The comparison between morning and early afternoon data shows that the variability is higher in the morning and lower in the early afternoon.

3.1.2. Vertical Profiles

Vertical profiles of temperature, relative humidity and specific humidity in the presbytery, as measured on 22 April 2022 (Figure 6), show very low variability. In the seven positions in this area, the vertical profile is relatively constant. The vertical homogeneity in temperature, relative humidity and specific humidity is a consequence of air well mixing.

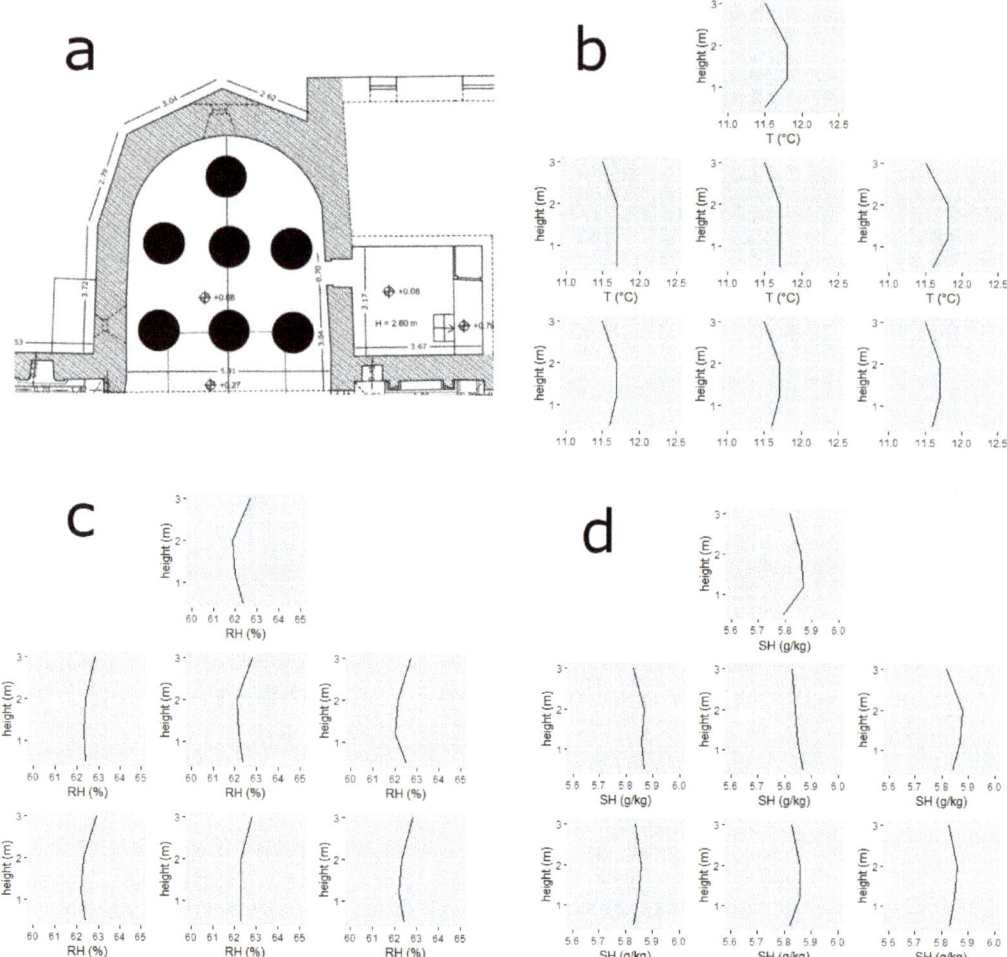

Figure 6. (**a**) Positions of the vertical profiles, (**b**) temperature profiles, (**c**) relative humidity profiles and (**d**) specific humidity profiles on 22 April 2022 at 14:30 UTC+1.

Fresco surface temperature was measured on 22 November 2021 and 22 April 2022 with an infrared thermometer. In both cases, the temperature values show a vertical profile with lower values at the ceiling and higher ones near the floor (Figure 7). Moreover, the walls exposed to the outside (left side in Figure 7) are characterized by lower temperature values in comparison with walls facing south or bordering the rectory (right side in Figure 7).

3.2. Continuous Monitoring

The monitoring activity involved the analysis of data collected inside the church at twelve sites (Figure 2b) and two sites outside the church, one below the portico and a second at the meteorological station of Colle San Vito.

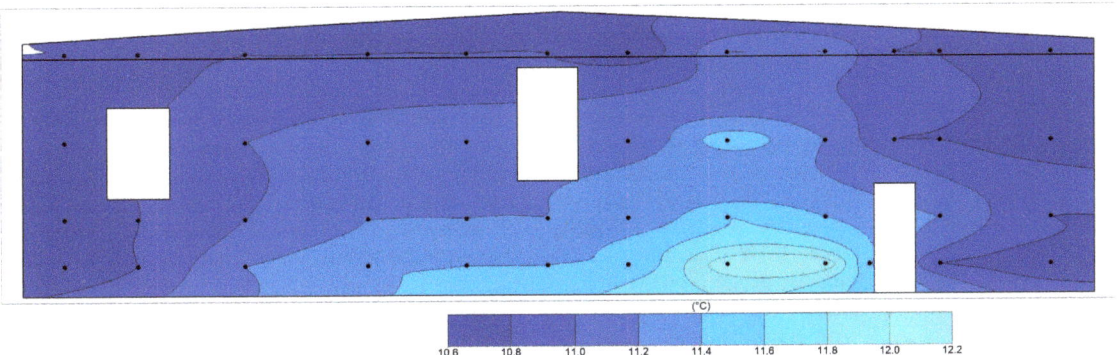

Figure 7. Apse wall temperature on 22 April 2022 at 14:30 UTC+1.

3.2.1. Apse Data Overview

In the apse, the temperature is modulated by seasonality, with the highest values in August and the lowest in February. In addition, it is influenced by meteorological events and, inside the church, by the presence of people during the weekly Sunday service. The overlapping of seasonality (time scale of one year), meteorological events (time scale of one week) and people during the services (time scale of one hour) is recognizable in data collected in the apse near the walls and in the central side (Figure 8a). The air in the central part of the apse (crucifix top, green line in Figure 8a) shows more influence from people with respect to the air near the wall (orange line in Figure 8a). In fact, air temperature near the wall is influenced by the wall temperature itself, which varies slowly as a consequence of the walls' high thermal capacity.

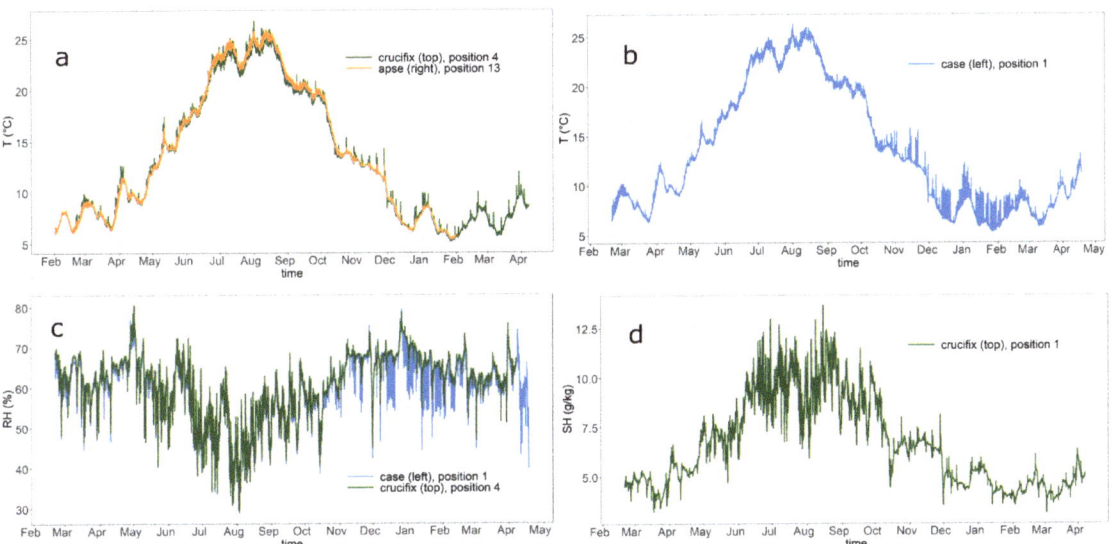

Figure 8. (**a**) Temperature at crucifix top (green line) and apse near wall right (orange line); (**b**) temperature on the case (blue line); (**c**) relative humidity at the crucifix top (green line) and on the case (blue line); (**d**) specific humidity at crucifix top.

The sensors over the glass showcase near the central window of the apse measure temperature values similar to those recorded near the crucifix with, in addition, a number of spikes from October to March (Figure 8b).

Relative humidity can vary inside the church in the range of 30–80%. The sensors on the glass showcases recorded negative spikes of about 12% in conjunction with temperature peaks (Figure 8c).

Specific humidity has been computed from temperature and relative humidity (Figure 8d). The time series show, as expected, high values in summer and low values in winter due to seasonality. In addition, variations due to meteorological events and increased values during the weekly services are present. In fact, respiration and transpiration produce moisture whose quantity depends on the environmental temperature and the physical activity of people [1]. In the specific humidity time series collected on the showcase, the spikes that appear in temperature and relative time series are not present.

3.2.2. Spikes Analysis

As mentioned in the previous section, some spikes in temperature and relative humidity were monitored at the showcase position. An inspection of data collected in the main nave reveals the presence of spikes in some positions during several periods in the year. In particular, abrupt variations in temperature and relative humidity were measured at the left balustrade and on the choir surface both on the left side and right side.

On the showcase, the spikes occurred during winter (from September to April) at the same day time (12:00–13:00 UTC+1) with a variation of temperature of about 3 °C and of relative humidity of about 12% and a time rising of more than 1 h (Figure 9a,b). As mentioned before, the specific humidity did not vary.

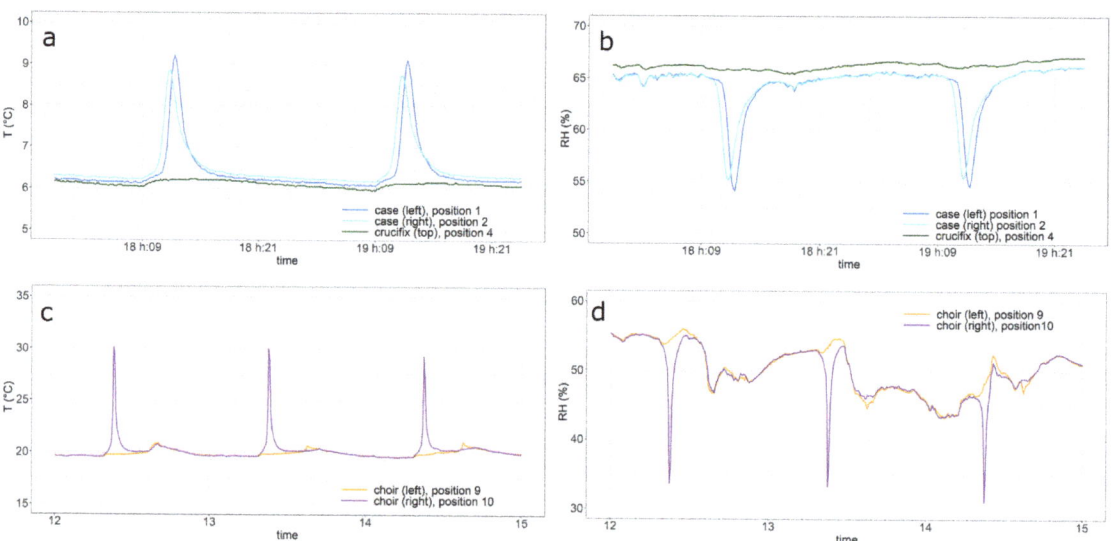

Figure 9. (a) Temperature on the case and at crucifix top on 18 and 19 January 2022; (b) relative humidity on the case and at crucifix top on 18 and 19 January 2022; (c) temperature at the choir from 12 to 14 September 2021; (d) relative humidity at the choir from 12 to 14 September 2021.

Near the left balustrade, spikes in temperature and relative humidity are recorded during two periods: from 19 August to 10 September and from 3 to 15 April at about the same day time (from 07:30 to 07:50 UTC+1). The intensities of peaks were, respectively, in the range of 2.1 °C to 3.0 °C and 8% to 10% with a rising time of 20 min.

On the right and left side of the choir, the peaks in temperature and relative humidity were detected in four periods during the year that are not always overlapping. As shown in Figure 9c,d, in the period 12–14 September, some peaks occurred in temperature and relative humidity on the right choir side but not on the left side. The daytime peak was always in the afternoon (from 14:00 to 15:10 UTC+1) on the choir's left side and in the morning (from 09:00 to 09:50 UTC+1) on the choir's right side with a rising time of 20 min. The intensity of these temperature peaks ranged from 6 °C to 12 °C, and the intensity of relative humidity peaks was from 15% to 30%. At the same time, the specific humidity values were subject to small positive peaks.

The timing in temperature and relative humidity peaks and the church window positions suggest that the peaks can be due to the presence of direct radiation on the surfaces. In particular, the glass showcase is exposed to solar radiation through the central apse window in the central day hours in winter when the sun is low on the horizon. The right size of balustrade and the choir are exposed to solar radiation in the morning or in the afternoon on different months of the year as a consequence of the different positions of windows on the lateral side of the apse and the main nave.

3.2.3. Psychrometric Diagrams

The microclimatic conditions at the different positions are summarized in psychrometric diagrams as shown in Figure 10, where the point cloud represents the air conditions and the red perimeter identifies the values permitted by the UNI standard [3] for frescoes (temperature ranges from 10 °C to 24 °C and relative humidity ranges from 55% to 65%). The comparison between the microclimatic conditions at the crucifix top (Figure 10c) and the same physical quantities outside the church below the external portico and at the meteorological station of Colle San Vito (Figure 10a,b) shows the church's ability to limit extreme temperature values and prevent cases of air saturation.

The ancient stone walls, with their great thermal capacity, adsorb heat in the warm season and release it in the cold one, resulting in moderate excursions in temperature under the external portico with respect to ones at the meteorological station of Colle San Vito, and even more moderate temperature values inside the church.

Likewise, the church building limits the exchange of moisture. Under the portico, the saturation air condition never occurred, and inside the church, relative and specific humidity were, respectively, never greater than 80% and 12 g/kg. The frescoes under the portico and the Madonna and Child in the lunette on the main church door were therefore not affected by condensation events on the surfaces.

Moreover, the church building limits the microclimatic states close to the intervals recommended by the norms. In fact, in Figure 10c, the points representing the microclimatic condition inside the church are concentrated close to the perimeter that identifies the conditions recommended by the UNI 10828 standard [3].

The dew point spread is defined as the difference between temperature and dew point temperature; it is an important physical quantity in microclimatic analysis as it gives information about saturation danger. If the dew point spread is close to 0 °C or lower, condensation can occur; otherwise, its value gives information about the distance from the saturation condition. In the presbytery, the air condition is always far from condensation (Figure 11a), whereas under the portico (Figure 11b), dew point spread is lower but it is always positive.

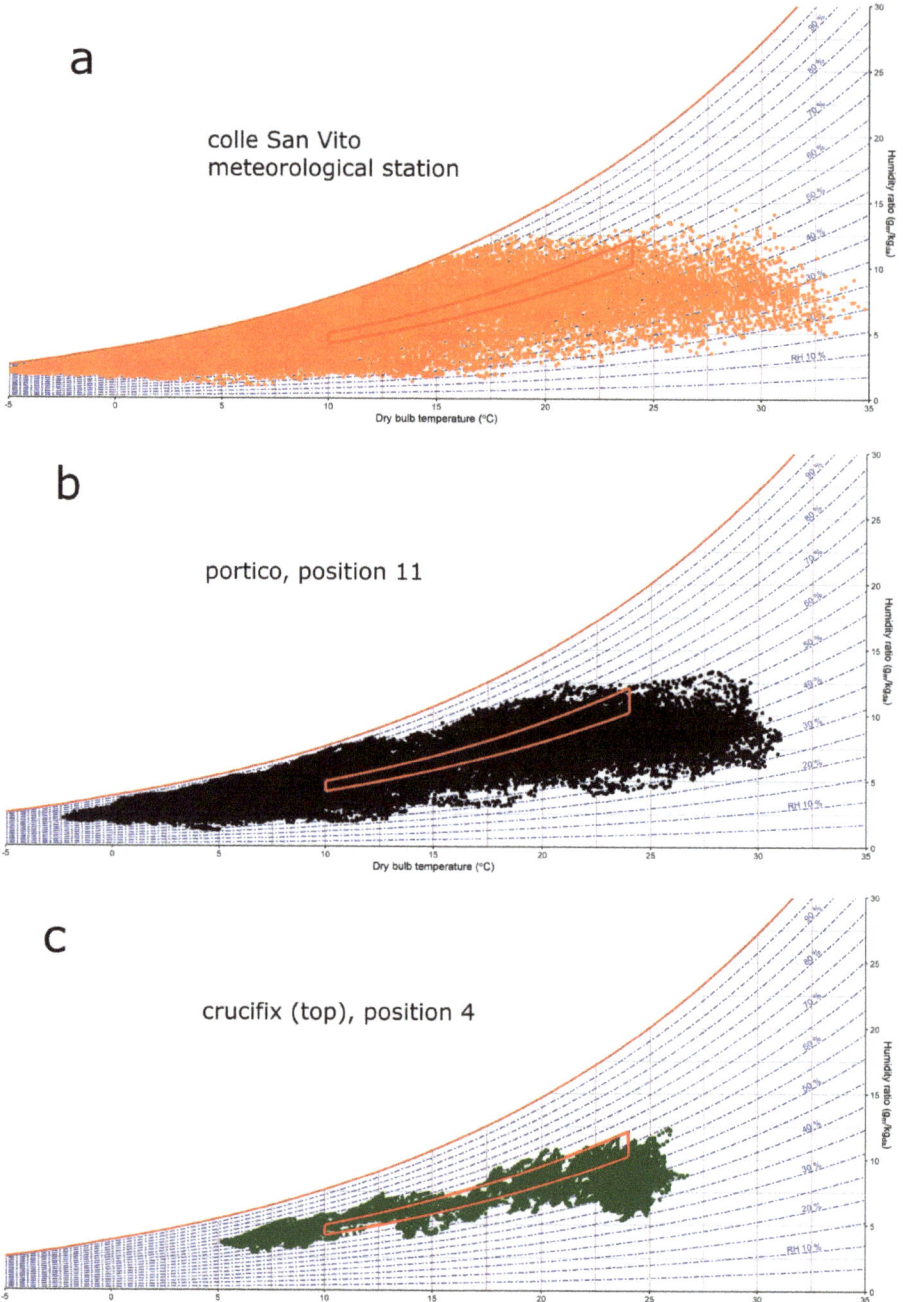

Figure 10. Psychrometric chart (**a**) at San Vito meteorological station, (**b**) outside San Panfilo church below the external portico and (**c**) inside the church at the crucifix top. The red perimeter identifies the conditions permitted by the UNI 10828 standard.

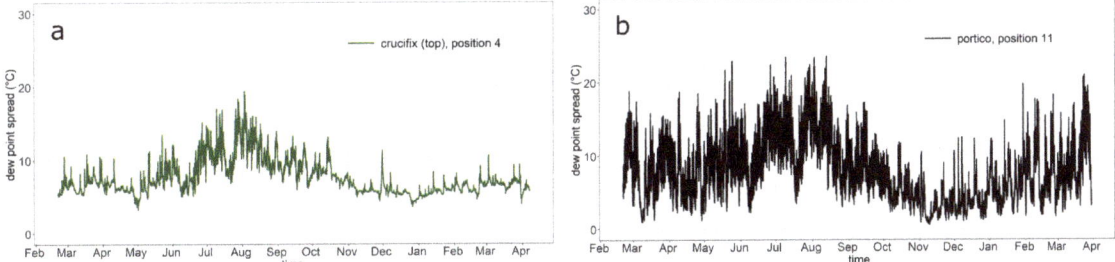

Figure 11. (**a**) Dew point spread at crucifix top and (**b**) dew point spread below the external portico.

3.2.4. Daily Averages

In order to compare indoor and outdoor temperature, the daily averages have been computed during the whole campaign period (Figure 12). The measured data at the crucifix top are chosen as representative of the indoor conditions, measured data under the external portico represent the environment near the church building and the measurements at the meteorological station of Colle San Vito describe the outdoor conditions. The mean daily excursion in temperature at the meteorological station of Colle San Vito in unperturbed conditions is 10.5 ± 3.5 °C, near the church it is 5.2 ± 2.0 °C and inside the church it is reduced to 0.8 ± 0.6 °C. Outdoor daily temperatures near the church building and at the meteorological station have the same trend, and, as expected, the temperature values measured under the portico are slightly higher than the ones at the meteorological station. Daily averaged temperature in the apse has lower fluctuations as the thermal capacity of the building is able to filter temperature variations due to meteorological events. In summer, the temperature values inside the church represent the averaged state of outdoor temperature, whereas in the cold season, it has higher values as a consequence of adsorbed heat during the summer.

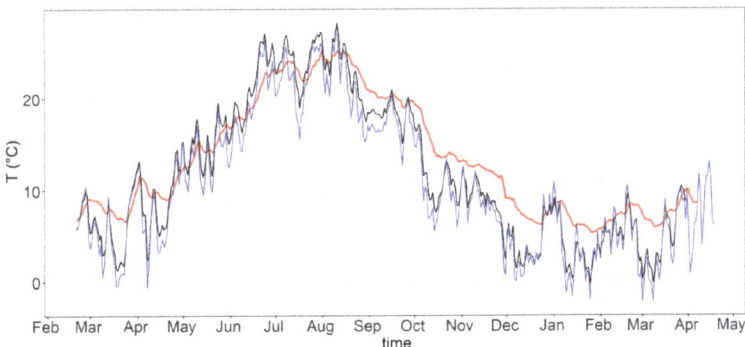

Figure 12. Daily temperature inside the church (red), outside below the external portico (black) and at the S. Vito meteorological station (blue).

Cross-correlation between daily temperature under the portico and inside the church reveals that the maximum correlation (R = 0.97) is reached with a delay time of 3 days.

3.2.5. Monthly Averages

The average values of temperature were computed for each month as mean of daily data for both outdoor and indoor (Figure 13). The annual trends show that, in general, the temperature below the portico is always higher than one at the meteorological station; moreover, in general, indoor temperature in the apse is higher than the temperature below the portico except during May, June and July. The scatter diagrams (Figure 13b,c) between indoor and outdoor monthly temperatures show the typical behavior of ancient and thick walls [8] absorbing heat during the summer and releasing it during the winter. In San Panfilo church, the internal heat sources are the people and the lighting system.

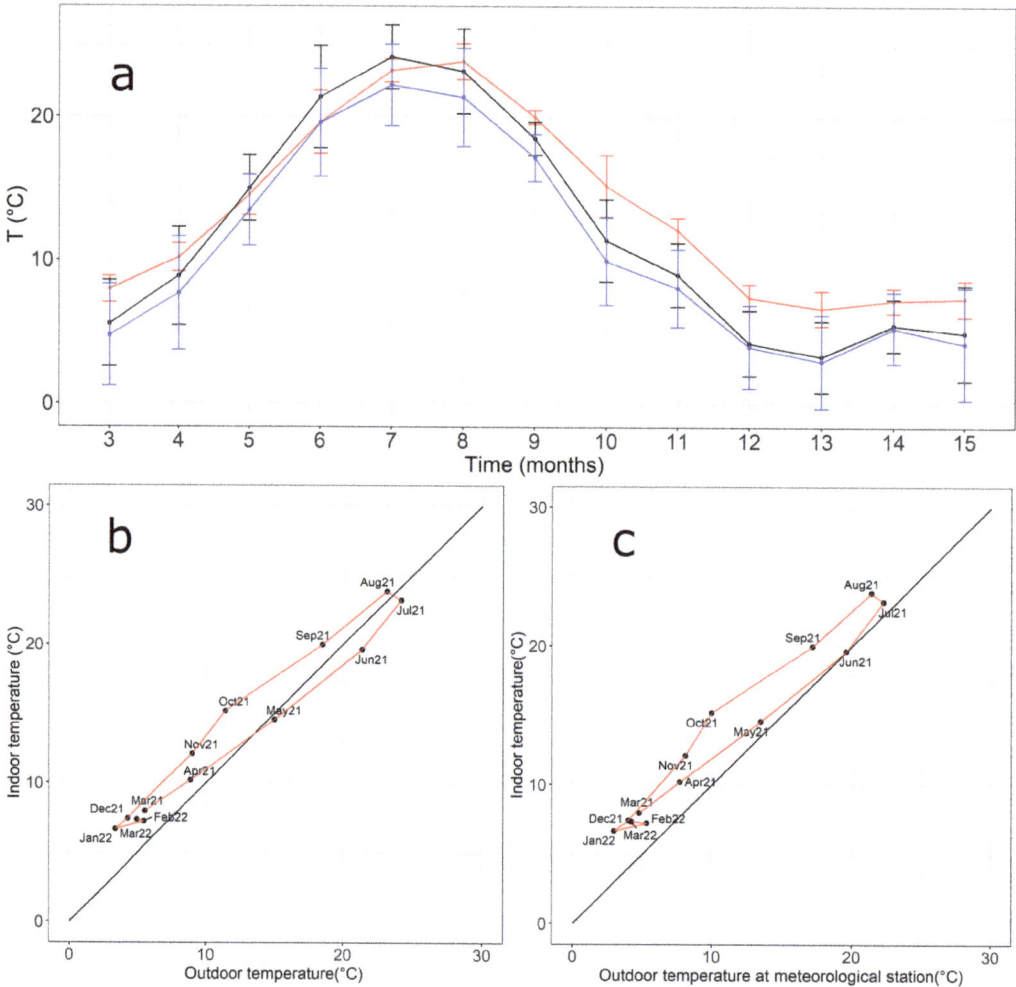

Figure 13. (a) Temperature monthly averages inside the church (red), below the external portico (black) and at the S. Vito meteorological station (blue); (b) scatter diagram with inside and outside (portico) monthly temperature; (c) scatter diagram between inside and outside (meteorological station) monthly temperature.

3.3. Historical Climate

The EN 15757 standard [4] provides a methodology to individuate the historical climate of works of art and to identify the risky conditions. It considers a dataset of relative humidity collected at 15 min sampling time during a period of 13 months. The distribution of fluctuations of relative humidity data with respect to the centered moving average (30-day time window) are computed, and the 7th and 93rd percentile are individuated. The standard establishes a safe band between the 7th and 93rd percentiles and a risky area outside this interval.

The historical climate is useful information for maintaining works of art in favorable microclimatic conditions for conservation. In particular, the historical climate should be preserved during restoration projects, avoiding abrupt fluctuations.

In order to characterize the historical climate inside the church and, in particular, in the apse, the EN 15757 procedure was applied to the time series collected at the crucifix top (Figure 14a,b). Unfortunately, the sensors near the walls (positions 13 and 14 in Figure 2b and Table 1), where Saturnino Gatti painted the frescoes, measured relative humidity for 12 months and at the sampling time of 60 min, so the methodology cannot be strictly applied. If the procedure is applied in these non-standard conditions, the results in Figure 14b–d are obtained. The two trends in the center of the apse and near the walls are similar, with slightly more variable relative humidity data near the walls with respect to the center of apse, as a consequence of the exchange in humidity with the walls. The identified historical climate in the apse could be useful for restorers during the next restoration work.

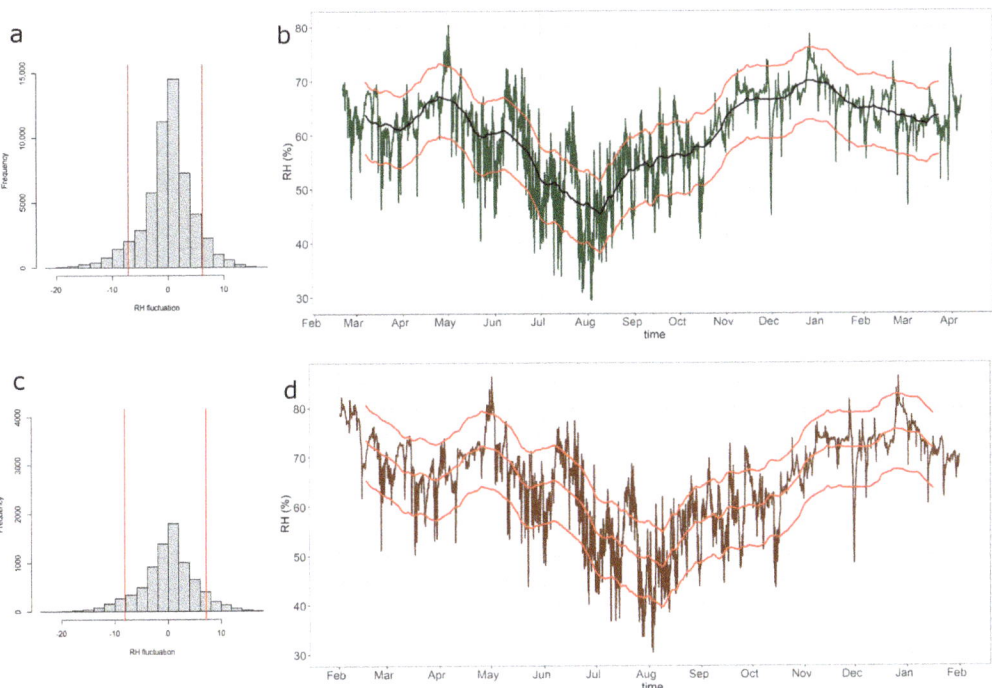

Figure 14. Historical relative humidity variability according to EN 15757 standard, in the figures red lines and curves show the 7th and 93rd percentiles; (**a**) distribution of relative humidity fluctuations at crucifix top site; (**b**) relative humidity variability at the crucifix top side (green line); (**c**) distribution of relative humidity fluctuations in the left side of the apse; (**d**) relative humidity variability at the left side of the apse (brown line).

4. Conclusions

The microclimate inside the church of San Panfilo in Tornimparte has been monitored in a measurement campaign that lasted 15 months from February 2021 to April 2022. The analysis was performed with intensive measurements on four days during the year and with continuous measurements at twelve sites inside the church and two sites outside during the whole period. Temperature and relative humidity were measured, whereas sensible humidity, dew point temperature and dew point spread were computed.

During the whole monitoring campaign, the church was routinely used by the people of Tornimparte for Sunday services and occasionally for funerals and weddings, so the environmental conditions were influenced by the presence of people, whereas the heating system did not modify them as it was turned off.

The intensive measurement data show horizontal homogeneity in temperature, relative and specific humidity, with one exception in summer, when more moisture is revealed near the main door on the left, likely due to moisture infiltrations in the wall. In the presbytery, the air is well mixed, and the wall temperature is lower near the roof and higher on the southern side in correspondence with the adjacent rectory.

The continuous data analysis provides evidence that direct solar radiation through windows on surfaces increases air temperature and decreases air relative humidity. The surfaces in the presbytery that are exposed to solar radiation through the central and lateral windows may increase their temperature; in particular, on the glass showcase on the presbytery, the temperature variation is about 3 °C in about one hour. Temperature variations inside the showcase (not measured), as a consequence of the greenhouse effect, are expected to be greater. The authors suggest to consider the eventuality of moving the glass showcase containing the silver processional crucifix to a protected position from direct solar radiation.

In the church, there is no danger of condensation as the dew point spread is always greater than 3 °C; below the external portico, condensation conditions are more likely, although they have never occurred throughout the year.

Church walls, as a consequence of their thermal capacity, reduce the variability in temperature, relative humidity and specific humidity. The measured values are close to the recommended ones by the UNI 10828 standard [3] for frescoes. Thick and massive walls reduce daily temperature fluctuations and introduce a delay time of about three days.

The comparison between average monthly indoor and outdoor temperatures shows that air inside the church is warmer than outside in the months from August to April, while it is characterized by lower temperatures in the months of May, June and July. This behavior is typical of ancient buildings with thick walls and has been identified in massive historical churches [8].

The application of the EN 15757 standard [4] allows to supply to the restorers the range of historical variability of relative humidity that should be respected during the restoration project.

Author Contributions: Conceptualization, S.F., D.B., A.G., L.L., C.P. and A.R.; methodology, S.F., D.B. and A.G.; software, S.F.; formal analysis, S.F.; investigation, S.F., D.B., A.G. and C.P.; data curation, S.F., D.B. and C.P.; writing—original draft preparation, S.F.; writing—review and editing, S.F., D.B., A.G., L.L., C.P. and A.R.; visualization, S.F.; project administration, A.R.; funding acquisition, A.R. All authors have read and agreed to the published version of the manuscript.

Funding: This work was performed in the framework of the Project Tornimparte—"Archeometric investigation of the pictorial cycle of Saturnino Gatti in Tornimparte (AQ, Italy)" sponsored in 2021 by the Italian Association of Archeometry AIAR (www.associazioneaiar.com).

Institutional Review Board Statement: Not applicable.

Informed Consent Statement: Not applicable.

Data Availability Statement: The data that support the findings of this study are not public; however, they are available for scientific purpose from the corresponding author.

Acknowledgments: The authors thank Domenico Fusari, President of Tornimparte Pro Loco, for his helpfulness and kindness; the parish priest and the deacon of the church of San Panfilo for the church availability and for their patience during the measurement campaign; Anna Galli, Letizia Bonizzoni and Simone Caglio for positioning the sensors during the COVID-19 lockdown; the Centro Funzionale and Ufficio Idrografico of Regione Abruzzo and the person of Giancarlo Boscaino for the availability of the dataset collected at the meteorological station of Colle San Vito.

Conflicts of Interest: The authors declare no conflict of interest.

References

1. Camuffo, D. *Microclimate for Cultural Heritage*, 3rd ed.; Elsevier: Amsterdam, The Netherlands, 2019; pp. 1–552.
2. Bernardi, A. *Microclimate Inside Cultural Heritage Buildings*; Ed il Prato: Padova, Italy, 2008; pp. 1–171.
3. UNI 10829; Works of Art of Historical Importance–Ambient Conditions for the Conservation-Measurements and Analysis. UNI Standard Ente Nazionale Italiano di Unificazione: Milano, Italy, 1999.
4. EN 15757; Conservation of Cultural Property-Specifications for Temperature and Relative Humidity to Limit Climate-induced Mechanical Damage in Organic Hygroscopic Materials. UNE-EN, AENOR: Madrid, Spain, 2010.
5. Schito, E.; Testi, D.; Grassi, W. A Proposal for New Microclimate Indexes for the Evaluation of Indoor Air Quality in Museums. *Buildings* **2016**, *6*, 41. [CrossRef]
6. Ferrarese, S.; Bertoni, D.; Dentis, V.; Gena, L.; Leone, M.; Rinaudo, M. Microclimatic analysis in the Museum of Physics, University of Turin, Italy: A case-study*. *Eur. Phys. J. Plus* **2018**, *133*, 538. [CrossRef]
7. Ferrarese, S.; Bertoni, D.; Golzio, A. An index for the evaluation of microclimatic conditions inside museum showcases. *Eur. Phys. J. Plus* **2022**, *137*, 1376. [CrossRef]
8. Camuffo, D.; Sturaro, G.; Valentino, A. Thermodinamic exchanges between the external boundary layer and the indoor microclimate at the Basilica of Santa Maria Maggiore, Rome, Italy: The problem of conservation of ancient works of art. *Bound. Layer Meteorol.* **1999**, *92*, 243–262. [CrossRef]
9. Bernardi, A.; Todorov, V.; Hiristova, J. Microclimatic analysis in St. Stephan's church, Nessebar, Bulgaria after interventions for the conservation of frescoes. *J. Cult. Herit.* **2000**, *1*, 281–286. [CrossRef]
10. Baggio, P.; Bonacina, C.; Romagnoni, P.; Stevan, A.G. Microclimate analysis of the Scrovegni Chapel in Padua. *Sud. Conserv.* **2004**, *49*, 161–176.
11. Andretta, M.; Coppola, F.; Seccia, L. Investigation on the interaction between the outdoor environment and the indoor microclimate of a historical library. *J. Cult. Herit.* **2016**, *17*, 75–86. [CrossRef]
12. Aste, N.; Adhikari, R.; Buzzetti, M.; Della Torre, S.; Del Pero, C.; Leonforte, F. Microclimatic monitoring of the Duomo (Milan Cathedral): Risks-based analysis for the conservation of its cultural heritage. *Build. Environ.* **2018**, *148*, 240–257. [CrossRef]
13. Lanteri, L.; Lo Monaco, A.; Pelosi, C. The relevance of monitoring the microclimate in museums. The case of Colle del Duomo in Viterbo. *Eur. J. Sci. Theol.* **2020**, *16*, 181–191.
14. Cataldo, R.; De Donno, A.; De Nunzio, G.; Leucci, G.; Nuzzo, L.; Siviero, S. Integrated methods for analysis of deterioration of cultural heritage: The Crypt of "Cattedrale di Otranto". *J. Cult. Herit.* **2005**, *6*, 29–38. [CrossRef]
15. Varas-Muriel, M.; Fort, R.; Martínez-Garrido, M.; Zornoza-Indart, A.; López-Arce, P. Fluctuations in the indoor environment in Spanish rural churches and their effects on heritage conservation: Hygro-thermal and CO_2 conditions monitoring. *Build. Environ.* **2014**, *82*, 97–109. [CrossRef]
16. Sileo, M.; Gizzi, F.T.; Masini, N. Low cost monitoring approach for the conservation of frescoes: The crypt of St. Francesco d'Assisi in Irsina (Basilicata, Southern Italy). *J. Cult. Herit.* **2017**, *23*, 89–99. [CrossRef]
17. Mannetti, T.R.; Chelli, N.; Vecchioli, G. *Saturnino Gatti nella Chiesa di San Panfilo a Tornimparte*; Edizioni del Gallo Cedrone: L'Aquila, Italy, 1992; pp. 1–72.
18. Arbace, L.; Di Paolo, G. *I Volti Dell'anima, Saturnino Gatti: Vita e Opere Di Un Artista Del Rinascimento*; De Siena Editore, Ed.; De Siena: Pescara, Italy, 2012.
19. Galli, A.; Alberghina, M.F.; Re, A.; Magrini, D.; Grifa, C.; Ponterio, R.C.; La Russa, M.F. Special Issue: Results of the II National Research project of AIAr: Archaeometric study of the frescoes by Saturnino Gatti and workshop at the church of San Panfilo in Tornimparte (AQ, Italy). *Appl. Sci.* **2023**; *to be submitted*.
20. Bolton, D. The Computation of Equivalent Potential Temperature. *Mon. Weather Rev.* **1980**, *108*, 1046–1053. [CrossRef]

Disclaimer/Publisher's Note: The statements, opinions and data contained in all publications are solely those of the individual author(s) and contributor(s) and not of MDPI and/or the editor(s). MDPI and/or the editor(s) disclaim responsibility for any injury to people or property resulting from any ideas, methods, instructions or products referred to in the content.

Article

Application of Sonic, Hygrometric Tests and Infrared Thermography for Diagnostic Investigations of Wall Paintings in St. Panfilo's Church

Sara Calandra [1,2], Irene Centauro [2], Stefano Laureti [3], Marco Ricci [3,*], Teresa Salvatici [2] and Stefano Sfarra [4]

1. Department of Chemistry, University of Florence, 50019 Florence, Italy; sara.calandra@unifi.it
2. Department of Earth Sciences, University of Florence, 50121 Florence, Italy; irene.centauro@unifi.it (I.C.); teresa.salvatici@unifi.it (T.S.)
3. Department of Informatics, Modelling, Electronics and System Engineering, University of Calabria, 87036 Rende, Italy; stefano.laureti@unical.it
4. Department of Industrial and Information Engineering and Economics, University of L'Aquila, 67100 L'Aquila, Italy; stefano.sfarra@univaq.it
* Correspondence: marco.ricci@unical.it

Featured Application: Non-destructive testing techniques, namely sonic pulse velocity, hygrometric tests, and infrared thermography, were employed for inspecting late XV-century frescoes to evaluate the state of conservation of the plaster-masonry structure.

Abstract: Prior to restoration work, the frescoes created at the end of the XV century by the painter Saturnino Gatti (1463–1518) in the apse of the Church of St. Panfilo in Villagrande di Tornimparte (L'Aquila) were the subject of a thorough diagnostic study involving several tests, from in situ non-destructive analysis to laboratory micro-destructive analysis on the collected samples. In this paper, we report the application of the sonic pulse velocity test, hygrometric tests, and infrared thermography to assess the state of conservation of the frescoes, i.e., the combined system of plaster and wall support. The complete analysis of the frescoes' state of conservation revealed significant insights. The integrity of the plaster was evaluated through sonic pulse velocity tests, which highlighted several areas of detachment or degradation phenomena. Hygrometric analysis described humidity variations, particularly near the boundary between the conch area and the church naves. Passive infrared thermography detected temperature inhomogeneities, emphasizing differences in the wall texture and the masonry structure. Moreover, by comparing sonic pulse velocity and passive thermography images, a certain degree of correlation between hot areas and slow areas in the presence of possible detachments was noticed. In addition, pulse-compression active thermography was applied in a few spots, and for the first time, to the best of our knowledge, the virtual wave concept was applied to the cultural heritage field. This strategy helps in better associating anomalies with depth. The measurement campaign was part of a research project conducted by members of the Italian Association of Archaeometry (AIAr), and the results were compared and integrated with those of other non-destructive and analytical methods.

Keywords: frescoes; non-destructive testing; heritage science; sonic pulse velocity; hygrometric tests; infrared thermography

Citation: Calandra, S.; Centauro, I.; Laureti, S.; Ricci, M.; Salvatici, T.; Sfarra, S. Application of Sonic, Hygrometric Tests and Infrared Thermography for Diagnostic Investigations of Wall Paintings in St. Panfilo's Church. *Appl. Sci.* **2023**, *13*, 7026. https://doi.org/10.3390/app13127026

Academic Editor: Asterios Bakolas

Received: 4 May 2023
Revised: 2 June 2023
Accepted: 8 June 2023
Published: 11 June 2023

Copyright: © 2023 by the authors. Licensee MDPI, Basel, Switzerland. This article is an open access article distributed under the terms and conditions of the Creative Commons Attribution (CC BY) license (https:// creativecommons.org/licenses/by/ 4.0/).

1. Introduction

Archaeometry analysis is increasingly used as a preliminary step for restoration interventions on cultural heritage (CH) items such as historic frescoes. Over the years, several non-destructive testing (NDT) and micro-destructive methods have been tailored, combined, and used to support restoration interventions [1,2].

The state of preservation of a fresco and the choice of the most suitable restoration procedure depends on the evaluation of many aspects, from the historical analysis of the artwork to the study of execution technique and materials used, and to the analysis of the support. The latter is generally a complex stratified system characterized by different constituent materials and construction techniques. Assessing the state of health and preservation of such structures is not straightforward, as many parameters are unknown—for example, the thickness and the number of the various layers, the materials used, etc.

Among the NDT techniques, a few can penetrate the outer layers, the carbonated surface, and the painting layer, revealing info about the rough coat layer and the support, although possible detachments, the presence of water, and high humidity values can represent sources of structural damage and degradation phenomena that can affect the outer layers.

Sonic pulse velocity (SPV) is a method that uses elastic compression waves generated by a mechanical impact on the surface of a sample to investigate the materials in depth [3–6]. In the case of good adhesion between the painting layer and the underneath, and between the various layers in general, the elastic waves travel faster from the source point to the receiver points. Conversely, detachments, voids, and discontinuities of the plaster can hamper the propagation of the elastic waves, resulting in a lower sound propagation velocity.

In addition, by integrating the results of SPV tests with those obtained from other techniques, detachments, microfractures or lesions, and inhomogeneous areas can be distinguished [7,8]. A high level of humidity or the presence of water are two other aspects that can influence the measured velocity. While on the one hand, the presence of water affects the sonic velocity, causing a general increase [9], on the other hand, this is the main cause of the degradation of frescoes, leading to a general decrease of velocity [10]. To distinguish between a low velocity due to moisture/water or by detachments/weak adhesion, complementary information must be acquired.

Even if the hygrometric test (HT) measures the moisture level of the outer layer, usually a certain degree of correlation between surface and subsurface moisture is present. The moisture maps can then be superimposed on the SPV ones looking to highlight anomalous regions. Further, the moisture maps give helpful indications for the conservation and restoration of the painting layer.

Another NDT technique that is frequently used to assess the presence of water ingress and the distribution of moisture is infrared thermography (IRT), both in passive (mainly) and in active modalities [11–13]. Unlike HT, passive IRT produces images that give qualitative information on the wall under analysis, if the boundary conditions allow [14].

It is possible to understand how, similarly to SPV, IRT is sensitive to moisture, detachments, voids, cracks, weak adhesion, buried structures, etc.

In general, in thermograms, such kinds of defects are detected as a rise in the local temperature, whereas there is a decrease in moisture level. Combining SPV, HT, and IRT can help distinguish between moisture and detachments, voids, weak adhesion, and other defects.

It should be noted that some of the above-mentioned types of defects cannot always be detected, as the signal-to-noise ratio (SNR) of passive thermography images relies on temperature gradients due to environmental sources (i.e., different sun exposure) and on variations in the emissivity of materials. Active IRT is thus used to increase the SNR by exciting the sample under testing with a controlled heating source. By active thermography, larger sensitivities are obtained, even if the range of inspection is smaller than in passive IRT, and further depth-dependent information can be extracted, which is helpful to study the stratigraphy of the fresco support. In this way, a more qualitative moisture analysis can be carried out [15,16].

In this paper, we report the combined application of SPV, HT, and IRT to the frescoes created by Saturnino Gatti and his workshop between the years of 1491 and 1494 in the apsidal chapel of St. Panfilo Church in Villagrande di Tornimparte (L'Aquila). These frescoes illustrate key moments of Jesus' Passion. Moreover, the ceiling shows images of God, Angels, and the Blessed in Paradise, and represent a precious example of Italian Renascence art.

The measurement campaign was part of a research project conducted by members of the Italian Association of Archaeometry (AIAr), and the results were compared and integrated with those of other non-destructive and analytical methods. The scheme methodology of data/image acquisitions was affected by strict restrictions due to laws linked to the COVID-19 pandemic. The research groups worked in a disjointed manner, although guided by a protocol that described the main activities to do and the timelines.

The results of other non-destructive and micro-destructive analyses on the St. Panfilo's wall will be discussed in other articles of the Special Issue "Results of the II National Research project of AIAr: archaeometry study of the frescoes by Saturnino Gatti and workshop at the church of St. Panfilo in Tornimparte (AQ, Italy)". For in-depth details on the aims of the project, see the introduction of the Special Issue [17].

The paper is organized as follows: in Section 2, the methods used to collect and process experimental data are reported for SPV, HT, and IRT; in Section 3, the results are summarized. In Section 4 the results are discussed and conclusions are drawn.

2. Methods

2.1. SPV and HT Data Collection and Imaging Procedure

The sonic investigations exploit propagation in the material of elastic compression waves generated by a short elasto-mechanical impact on the surface to investigate the conditions in which the material is in its interior or to localize any inhomogeneities, voids, and defects in the investigated section [3–6,18].

The SPV is dependent on the physical-mechanical characteristics of the investigated material. Internal defects, voids, and lower density result in lower velocity values, indicating a poor state of conservation. On the contrary, high velocities indicate good characteristics and high homogeneity of the material.

The test involves the generation of elastic waves by using an instrumented impact hammer with a frequency range of 20 Hz–20 kHz through a punctual mechanical impulse, which is then read by an accelerometer. The value obtained in a sonic test is the time-of-flight (ToF), which is the time interval taken by the wave to pass through the material investigated. From the ToF and the length of the path investigated (L), the velocity of the wave (v) is estimated.

The instrumentation used for this study is a Novasonic U5200 CSD of IMG Ultrasuoni Srl, consisting of a hammer with a load cell in the impact head, the source of the acoustic wave. The time–amplitude function of the force applied in the impact is recorded through a detection system directly connected to the hammer [3,7].

The sonic test is performed in indirect mode; see Figure 1a in which the hammer and the accelerometer are placed on the same face. SPV was performed according to a specific survey scheme on the same regular grid of points (Figure 1b), following a horizontal line. The mechanical impulse point is kept fixed, moving the accelerometer horizontally at progressively increasing distances from the hammer.

By collecting measurements in a regular grid of points, SPV maps can be visualized to identify possible critical areas. These SPV maps represent velocity distribution. From the reference velocity of the investigated material in a good state of preservation, it is possible to consider the state of degradation of the frescoed masonry by comparing zones at different velocities. Since the SPV test involves acquiring large data, a management system for data collection and analysis on-site has been designed, adapting it to the specific survey campaign [3] (Figure 1c). Moreover, collected data were processed through GIS software for the graphical representation of the velocity distribution maps.

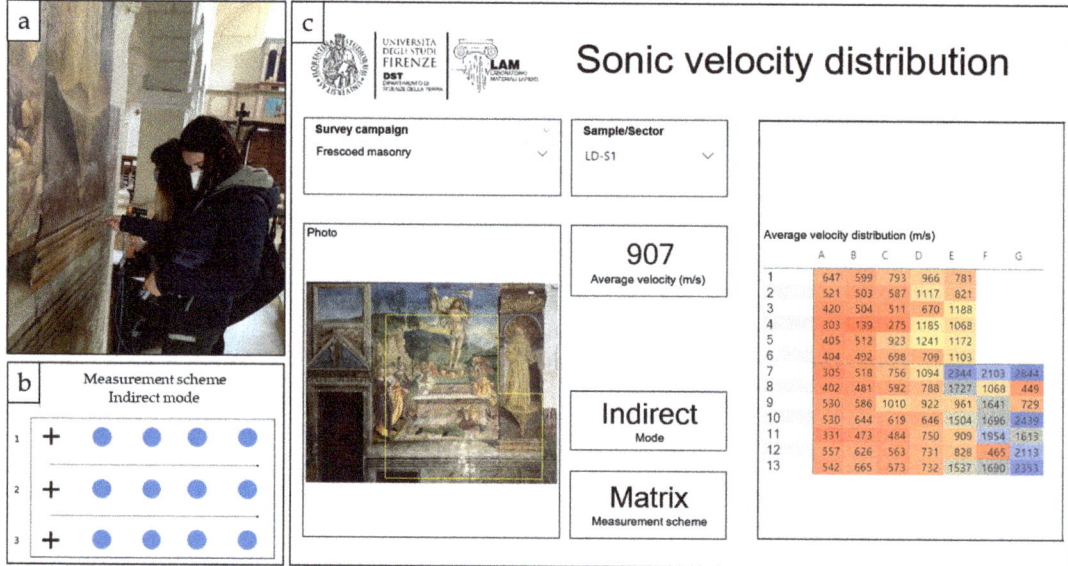

Figure 1. Sonic pulse velocity test: (**a**) on-site investigation, (**b**) measurement scheme performed in indirect mode. The + symbol indicates the hammer position, and the blue points represent the accelerometer position. (**c**) interactive report page to provide real-time verification of data and analysis.

Likewise, hygrometric tests (HT) of the area were performed punctually, following the SPV measurement scheme and data processing. HT was carried out using the instrument Protimeter MMS2 BLD8800 of Allemano Instruments, set on the "pin mode", to measure the surface moisture of the painting layer. The instrument measures the moisture content in WME% in non-conductive solid materials [19]. HT maps were generated by imaging the punctual values collected to highlight moisture distribution in the frescoed masonry. The environment measurements were conducted in the study area, with an indoor temperature of 7–12 °C and a relative humidity of 55–60%.

SPV and HT investigations were performed on the frescoed walls of the apsidal conch of St. Panfilo Church, selecting four areas (panels A, B, D, and E), widely described in [20].

2.2. IRT Data Collection and Imaging Procedure

The IRT tests were performed in both passive and active approaches (see [1]) by using a 640 × 480 thermal camera (7.5–13 µm, T660 FLIR System). Passive IRT (PIRT) was used to image frescoes' support walls and detect potential anomalies. PIRT uses infrared imaging to capture the natural temperature variations or mappings in materials and structures such as buildings [14]. Thanks to the rapidity of the passive IRT test, it was possible to collect images on the whole apse and structures close to it to evaluate the presence of possible areas with large inhomogeneities. For all the acquisitions, the visible RGB image of the area was also collected and superimposed to the thermogram to help interpret the thermal imprints; see Figure 2. During PIRT acquisitions, the thermal camera was handheld, and the target focused from a distance of around 6 m in a 4 °C and 55–60% RH environment.

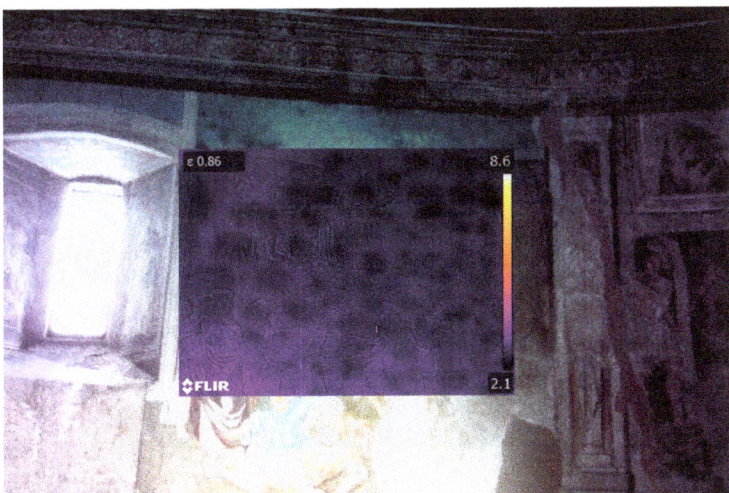

Figure 2. Example of passive thermography analysis used to identify the support wall texture. Stones with irregular shapes can be distinguished behind the fresco.

The active IRT was applied instead to some specific areas to acquire information at different depths. Two details of the frescoes' cycle were analyzed, as reported in the following section. The active IRT tests implemented exploited the pulse-compression thermography (PuCT) procedure introduced in [21] for the analysis of two Renaissance panel paintings and then further developed and applied to a variety of historical artworks and mock-ups; see, for example [22,23]. The peculiarity of the PuCT procedure is combining a long excitation time, such as in lock-in thermography, with the information content of pulsed thermography. This is made possible by using a pseudo-noise (PN) heat source (i.e., a heat source that is switched on and off following a PN signal) and by applying the pulse-compression procedure to the sequence of thermograms collected during PN excitation. In particular, the temperature time-sequence of each pixel is correlated with the PN input signal. After the PuCT, the results of a virtual pulsed thermography experiment are retrieved, and they can be further analyzed with standard thermography tools. In this way, a high value of SNR can be obtained by using even low-power excitations. In this work, an 8-LED system with a total electrical power of 400W was used to heat portions of the frescoes. The PN signal modulating the LED system was generated starting from a Legendre sequence of 31 bits, with a bit duration of 10s [22]. The overall duration of active thermography acquisitions was 620 s, i.e., approximately 10 min, and the thermogram acquisition rate was 2 frames per second, so each thermal sequence consisted of 1040 images.

3. Results

3.1. SPV and HT Results

The SPV and HT tests were performed on the frescoed walls of the apsidal conch of St. Panfilo's Church, selecting four areas characterized by fractures, infill of the underlying masonry, and near openings (doors and windows). For each panel, the SPV results are compared with the hygrometric one through maps obtained from data processing with GIS software. The point values have been interpolated to provide a 2D false-color map of the sonic velocity and hygrometric value distribution according to the Spline function. The maps in false color emphasize the different properties of the masonry and are shown in Figure 3.

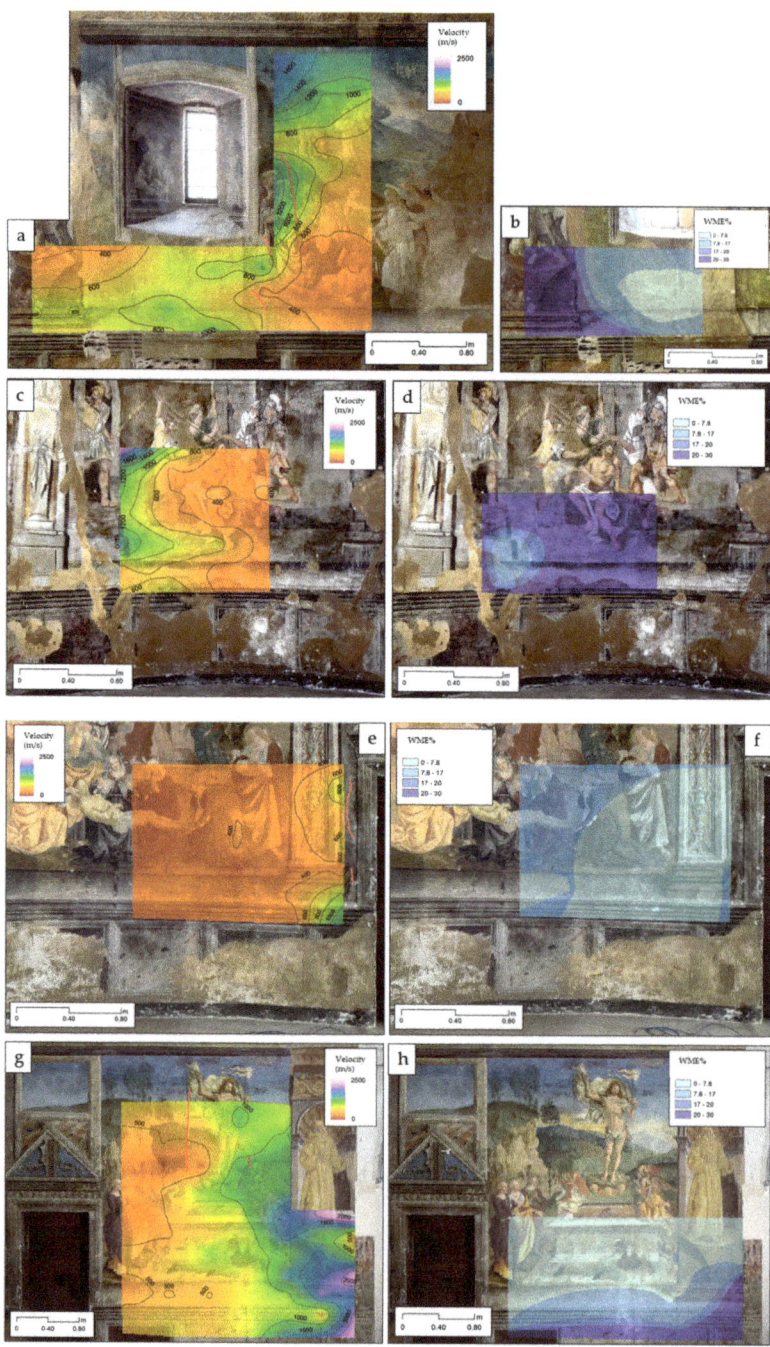

Figure 3. SPV and HT value distribution maps of (**a,b**) Panel A; (**c,d**) Panel B; (**e,f**) Panel D; and (**g,h**) Panel E. The plaster fractures are highlighted in red.

The sonic velocity of the walls ranges between 0 and 2500 m/s. In most cases, these values are related to the layer of painted plaster, thus being less representative of the underlying support masonry. False color maps highlight the different properties of the masonry. Higher velocity values correspond to well-cohesive plaster that is well-bonded to the masonry (areas from blue to purple in the maps), while lower velocity values indicate portions of detached or damaged plaster resulting from fractures (red areas).

In Figure 3a, the average velocity is about 700 m/s. The red area on the lower right shows lower velocities, which is due to both the detachment of the plaster and a large discontinuity that starts from the right edge of the window and cuts the masonry.

The HT values show that the humidity of the plaster increases towards the edge that delimits the conch area from that of the church naves (Figure 3b).

In Figure 3c, the average sonic velocity (v) is around 600 m/s, while in the right portion, average velocity values of about 370 m/s are recorded. These lower values are mainly due to a detached portion of the plaster that is located in the central portion of the investigated area. On the left side, however, where the plaster shows good cohesion to the masonry, the velocity is higher (average of 850 m/s).

In Figure 3d, the moisture values of the plaster are higher and more homogeneous (>20 WME%) than in the other areas investigated. This phenomenon is most likely due to the position of this area relative to the cardinal points; in fact, the wall is exposed to the sun only in the morning. The area corresponding to the detached portion of the plaster falls below 10 WME%, in accordance with SPV results.

In Figure 3e, the average sonic velocity is around 370 m/s. This area is affected by widely detached plaster. Moreover, a discontinuity is visible that divides the opening frame, probably made from stone, to the masonry and brings a variation of velocity; the right part has a higher velocity than the left side.

The results of the HT show that the moisture of the plaster increases going down (Figure 3f). In this area, the humidity value never exceeds 10 WME%, which can most likely be linked to the presence of numerous detached portions that allow air circulation. In addition, this wall of the conch communicates with the rooms of the rectory, so it is not directly exposed outside.

In Figure 3g, the left portion has lower sonic velocity values. On the right side instead, the velocity is higher (1500 m/s on average). This separation is found in the correspondence of a discontinuity in the coating plaster, as well as in the masonry itself (highlighted in red in Figure 3g). Figure 3h illustrates the HT values recorded, which indicate an increase in the moisture of the plaster in the lower right portion.

3.2. Passive and Active IRT Results

PIRT is useful to identify potential detachments, part of the wall structure, and a presumable humidity area. Figure 4 shows a part of the apse between two panels of the frescoes and precisely between the central one and that on the left (between panels A and B), which is the one with the worst degree of conservation and that was probably retouched in the past several times. A quite large and abrupt average measured temperature discontinuity between the central-right side of the apse and the left side can be seen, which was confirmed by the other images collected.

As mentioned above, this can be due to the position of the left side relative to the cardinal points, even if at the time of the measurement, the left side was exposed to the sun, but probably also to differences in the structure that affect the emissivity (e.g., materials, construction techniques, plaster characteristics, and later works such as the construction of the sacristy). This difference between the left and right sides is confirmed by the PIRT images shown in Figure 5 and collected over the same areas imaged with SPV and HT, reported in Figure 3. The subplots are named according to Figure 3.

Figure 4. Passive IRT image of the attachment point of panels A and B. An abrupt discontinuity in the average emitted radiation can be seen.

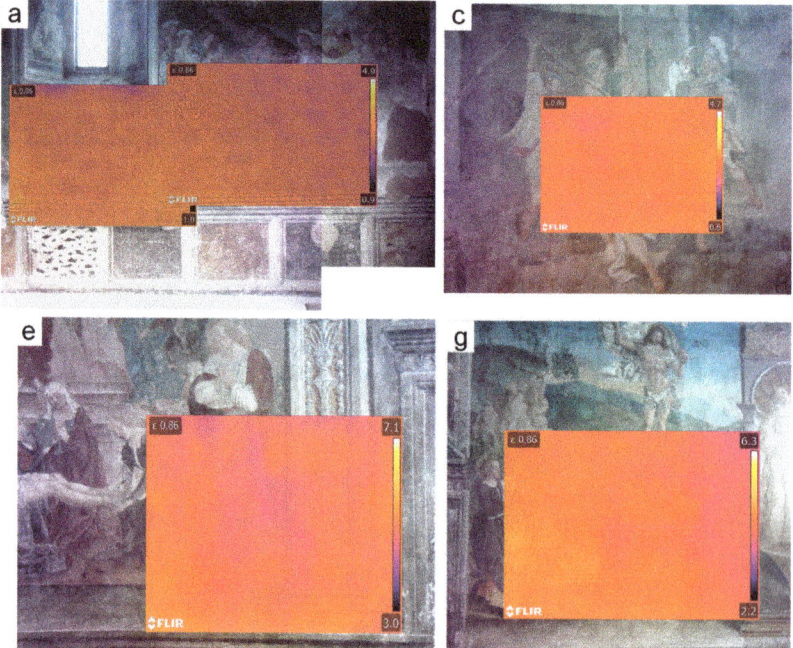

Figure 5. Passive IRT images of some frescoes' parts. The subplots (**a**,**c**,**e**,**g**) are named according to Figure 3.

In general, large temperature drops that could indicate large voids, anomalous moisture levels, or mass transfer (e.g., water) are not visible, even considering that the measurements were collected in the winter just after a huge snowfall (the church and all the area was covered by a thick layer of snow). There was also no evidence of severe inhomogeneities, moisture, or detachments, as the wall texture was visible in all the images, while large and long cracks were not. However, the wall texture was more visible on the left side of the apse, which is also less homogeneous with respect to the measured temperature.

The active IRT focused on two figures of the frescoes cycle. One is the figure that is going to be slaughtered in the second panel of the frescoes, starting from the left of the apse (panel A), and the other is the soldier lying at the foot of the tomb in the scene of the resurrection, which is in the first panel from the right (panel E). These two figures were chosen since passive IRT images evidenced some inhomogeneous spots on them, and further, for the soldier, the pictorial layer is significantly damaged, mainly scratched. In particular, the regions of interest (ROIs) were suggested by an art historian, Dr. Saverio Ricci, a renowned expert of the art master Saturnino Gatti.

The results of the PuCT experiments are reported in Figures 6 and 7. In particular, after the PuCT, virtual wave processing was applied [24]. These technique maps the time thermal signals, i.e., the intensity time trends of each pixel of IRT images, which are due to a diffusion process, into virtual signals associated to a propagation phenomenon, in which velocity is related to the thermal diffusivity. In this way, it is possible to estimate the depth corresponding to a specific image in the sequence after virtual wave application. This was made by assuming a thermal diffusivity of 0.3 mm^2/s. To the authors' knowledge, this is the first application of the virtual wave approach to the thermography inspection of CH.

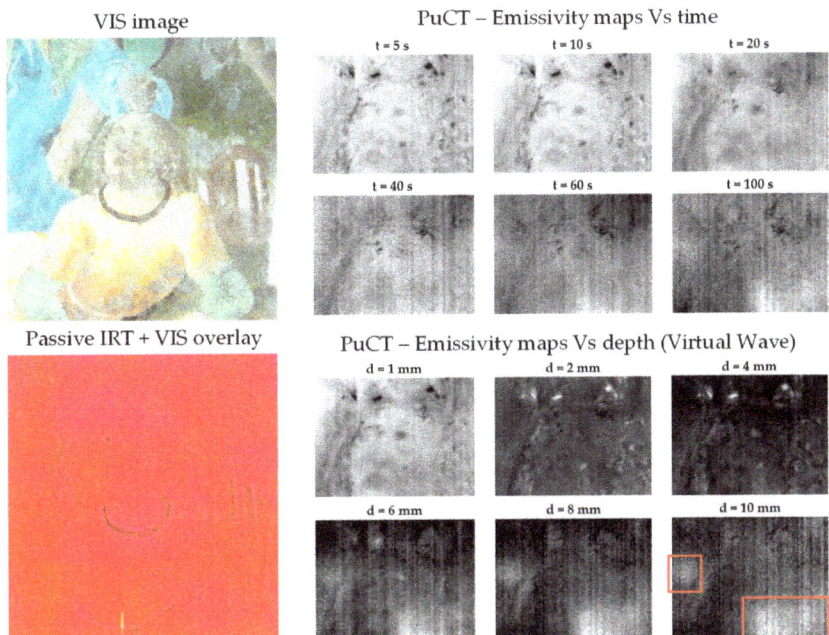

Figure 6. Results of PuCT and PuCT + virtual wave processing for the first area investigated. The red boxes in the bottom-right images indicate the shape of the bricks, which also barely visible in the passive IRT image (see Figure 5a). In the case of PuCT + virtual wave the bricks' shape appear from a certain value of depth and the maximum contrast is reached for a depth d = 10 mm.

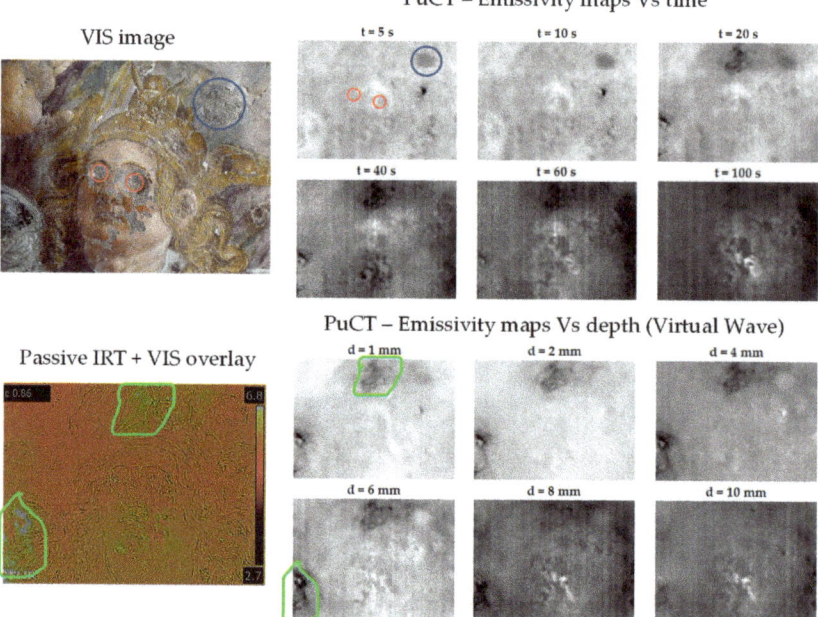

Figure 7. Results of PuCT and PuCT + virtual wave processing for the second area investigated. Blue and green curves highlight areas of specific interest. The blue circle indicates an area where a surface restoration intervention is visible. The thermal response is quite different indeed from the neighbor area especially at the initial times/depths, as expected from a surface inhomogeneity. The green curves indicate areas where there is no evidence of surface anomalies, but the thermal images highlight the presence of a thermal inhomogeneity under the surface.

Figure 6 reports the results for the first ROI. On the left, the VIS and the passive IRT images are shown, and the right shows some maps of the emissivity vs. time and emissivity vs. depth reconstructed after the PuCT and the PuCT + virtual wave processing.

In the time sequence of emissivity maps, the pictorial layer is maximally visible at the beginning, as expected. After some time, the maps are determined by the diffusion underneath the pictorial layer. However, the transition is smooth. After applying the virtual wave, the info is better associated with depth. In particular, the figure almost disappears beyond 2 mm, while other scratches and damages are visible up to 4 mm in depth. Further, the wall texture becomes clearly visible around 8–10 mm of depth; see the red rectangles in the right-bottom corner of Figure 6. This gives a measure of the plaster thickness and a better resolution than the passive IRT image can achieve. Note that a thermal diffusivity of 0.3 mm^2/s was assumed.

A similar analysis was done for the soldier figure, the second ROI, which is reported in Figure 7. The pictorial layer is barely visible in the emissivity maps due to the large degree of degradation. The wall texture is not clearly visible as well, as found by passive IRT, but instead, some interesting features were found.

For example, there are two hot spots (green marks in Figure 7) on the top of the soldier's forehead, just on the rabbit's left on the hat, and on the right shoulder. These hot spots are also visible in the passive IRT image, and with the virtual wave approach, their depth can be better evaluated.

4. Discussion and Conclusions

The integration of humidity, sonic velocity, and thermal maps provided complementary information that enabled the estimation of various critical factors such as detachments, rising damp, percolation, humidity, and other inhomogeneities, which represent typical decay phenomena in wall paintings. This comprehensive analysis allowed for the identification of potential critical areas, emphasizing the need for targeted conservation and restoration procedures for the wall paintings of St. Panfilo's Church.

In particular, the SPV is useful to diagnose masonry in a non-destructive way and deepen the knowledge of the construction characteristics, evaluate the state of conservation, and provide important information for effective restoration interventions of masonry.

The maps in SPV and HT enabled conservation problems (e.g. surface warping, painting layer detachments, and humidity distribution) to be evidenced and provided indications about different properties of the infill wall. These investigations aimed at assessing the consistency of the plaster coating and locating the possible presence of detachments, discontinuities, and voids.

The SPV and HT tests highlighted how both microclimatic factors and the different exposure of the masonry to solar radiation have a significant influence on the conservation state of the frescoes. Thermal stress can be decisive in generating detachments and fractures in the painted plaster, as evident from Figure 3a,b. This area near the window is characterized by humidity inhomogeneities, which also affect the plaster, as indicated by the highly variable SPV values. On the other hand, the areas whose external walls are less exposed to sunlight (Figure 3c,d) show higher and relatively homogeneous humidity values.

Regarding passive and active IRT, these techniques give information that can complement the SPV and HT maps.

By comparing the passive IRT and SPV images, a certain degree of correlation between hot areas and slow areas in the presence of possible detachments is noticed (see Figures 3 and 5). In particular, in subplots "e" and "g", the hottest areas correspond to low-sound velocity areas in SPV images where possible detachments can be found.

In many IRT images, the wall texture is visible, and in such areas, the presence of relevant detachments can be excluded. The PuCT active IRT, combined with virtual wave processing, can give detailed information about the plaster thickness and a better resolution than the well-known passive IRT approach (see Figure 6). All these advantages may help the restoration tangibly because the thermal imprints are read like a sort of tomography.

Based on the integrated use of SPV, HT, PIRT, and PuCT techniques, several observations can be made regarding the conservation status of the frescoes:

- plaster cohesion: several detached areas or damaged plaster resulting from fractures were evidenced;
- moisture distribution: the HT tests reveal variations in humidity levels across different areas. The moisture values are generally higher and more homogeneous in those regions potentially influenced by the position of the area relative to the sun exposure. Detached portions of plaster exhibit lower moisture values;
- temperature variations: the passive IRT analysis shows temperature discontinuities between different sections of the apse, indicating potential variations in wall texture and masonry structure. These differences may be attributed to factors such as materials, construction techniques, or later interventions;
- structural integrity of frescoes: the presence of fractures, discontinuities, and infill of the underlying masonry is evident from the SPV and PuCT results. These structural elements can affect the overall stability and conservation of the frescoes.

The integration of SPV, HT, and IRT results demonstrated a smart way to inspect artistic and architectural heritage, and this test represents an important starting point to deepen the knowledge of the structures and direct further and targeted investigations and restoration interventions.

The types of defects in a complex system such as frescoed masonry vary greatly. Therefore, the combination of the different non-destructive techniques proposed in this paper is helpful in overcoming the limitations of individual methods and achieving a more reliable diagnosis of the state of conservation of the wall paintings, allowing a comprehensive and multi-scale approach to the restoration problem.

This approach provides a comprehensive and multi-scale analysis of these complex structures, offering a more reliable and detailed understanding of their construction characteristics and conservation status. Additionally, the study highlights the significance of combining these techniques to overcome the limitations of individual methods and achieve a more comprehensive diagnosis, paving the way for targeted and effective restoration interventions.

Further steps, such as image fusion and correlation, are desirable in future applications to improve the accuracy of the tests and the evaluation of the conservation status.

Author Contributions: S.C., I.C. and T.S. equally contributed to the conceptualization, methodology, and experimental analysis of SPV and HT; S.L., M.R. and S.S. equally contributed to the conceptualization, methodology, and experimental analysis of IRT. All authors contributed to the writing. All authors have read and agreed to the published version of the manuscript.

Funding: This work was performed in the framework of the Project Tornimparte—"Archeometric investigation of the pictorial cycle of Saturnino Gatti in Tornimparte (AQ, Italy)" sponsored in 2021 by the Italian Association of Archeometry AIAR (www.associazioneaiar.com).

Institutional Review Board Statement: Not applicable.

Informed Consent Statement: Not applicable.

Acknowledgments: The authors want to thank the Proloco Tornimparte association, Giuseppe Spagnoli (University of L'Aquila), Saverio Ricci (Superintendence of Archaeology, Fine Arts and Landscape for the Provinces of L'Aquila and Teramo), and the mayor of the municipality of Tornimparte for their availability and support shown during the NDT campaigns. The authors are also grateful to LAM-DST-UNIFI for the technical and scientific support for the investigations, especially Carlo Alberto Garzonio.

Conflicts of Interest: The authors declare no conflict of interest.

References

1. Zuena, M.; Baroni, L.; Graziani, V.; Iorio, M.; Lins, S.; Ricci, M.A.; Ridolfi, S.; Ruggiero, L.; Tortora, L.; Valbonetti, L.; et al. The techniques and materials of a 16th century drawing by Giorgio Vasari: A multi-analytical investigation. *Microchem. J.* **2021**, *170*, 106757. [CrossRef]
2. Sandu, I.; Dima, A.; Sandu, I.G.; Luca, C.; Anca Sandu, I.C.; Sandu, A.V. Survey on behaviour of interventions for probota monastery indoor frescoes conservation under environmental factors influence. III. Correlations between thermal, hygroscopic and sonic parameters. *Environ. Eng. Manag. J. (EEMJ)* **2004**, *3*, 561–567. [CrossRef]
3. Centauro, I.; Calandra, S.; Salvatici, T.; Garzonio, C.A. System integration for masonry quality assessment: A complete solution applied to Sonic Velocity Test on historic buildings. In *The Future of Heritage Science and Technologies: Materials Science*; Springer International Publishing: Cham, Switzerland, 2022; pp. 213–226.
4. Coli, M.; Ciuffreda, A.L.; Donigaglia, T. Technical analysis of the masonry of the Bargello'Palace, Florence (Italy). *Appl. Sci.* **2022**, *12*, 2615. [CrossRef]
5. Luchin, G.; Ramos, L.F.; D'Amato, M. Sonic tomography for masonry walls characterization. *Int. J. Archit. Herit.* **2020**, *14*, 589–604. [CrossRef]
6. Binda, L.; Saisi, A.; Tiraboschi, C. Investigation procedures for the diagnosis of historic masonries. *Constr. Build. Mater.* **2000**, *14*, 199–233. [CrossRef]
7. Calandra, S.; Cardinali, V.; Centauro, I.; Ciuffreda, A.; Donigaglia, T.; Salvatici, T.; Tanganelli, M. *Integration of historical studies and ND techniques for the structural characterization of the masonry walls in Palazzo Vecchio, Florence, In Diagnosis of Heritage Buildings by Non-Destructive Techniques*; Tejedor, B., Bienvenido-Huertas, D., Eds.; Elsevier: Amsterdam, The Netherlands, 2023.
8. Solla, M.; Gonçalves, L.M.; Gonçalves, G.; Francisco, C.; Puente, I.; Providência, P.; Gaspar, F.; Rodrigues, H. A building information modeling approach to integrate geomatic data for the documentation and preservation of cultural heritage. *Remote Sens.* **2020**, *12*, 4028. [CrossRef]
9. Besharatinezhad, A.; Khodabandeh, M.A.; Rozgonyi-Boissinot, N.; Török, Á. The Effect of Water Saturation on the Ultrasonic Pulse Velocities of Different Stones. *Period. Polytech. Civ. Eng.* **2022**, *66*, 532–540. [CrossRef]
10. Miranda, L.; Cantini, L.; Guedes, J.; Binda, L.; Costa, A. Applications of sonic tests to masonry elements: Influence of joints on the propagation velocity of elastic waves. *J. Mater. Civ. Eng.* **2013**, *25*, 667–682. [CrossRef]

11. Garrido, I.; Lagüela, S.; Fang, Q.; Arias, P. Introduction of the combination of thermal fundamentals and Deep Learning for the automatic thermographic inspection of thermal bridges and water-related problems in infrastructures. *Quant. InfraRed Thermogr. J.* **2022**, 1–25. [CrossRef]
12. Rocha, J.H.A.; Santos, C.F.; Póvoas, V. Evaluation of the infrared thermography technique for capillarity moisture detection in buildings. *Procedia Struct. Integr.* **2018**, *11*, 107–113. [CrossRef]
13. Kim, C.; Park, G.; Jang, H.; Kim, E.J. Automated classification of thermal defects in the building envelope using thermal and visible images. *Quant. InfraRed Thermogr. J.* **2022**, 1–17. [CrossRef]
14. Larbi Youcef, M.H.A.; Feuillet, V.; Ibos, L.; Candau, Y. In situ quantitative diagnosis of insulated building walls using passive infrared thermography. *Quant. InfraRed Thermogr. J.* **2022**, *19*, 41–69. [CrossRef]
15. Davin, T.; Serio, B.; Guida, G.; Pina, V. Spatial resolution optimization of a cooling-down thermal imaging method to reveal hidden academic frescoes. *Int. J. Therm. Sci.* **2017**, *112*, 188–198. [CrossRef]
16. Cadelano, G.; Bison, P.; Bortolin, A.; Ferrarini, G.; Peron, F.; Girotto, M.; Volinia, M. Monitoring of historical frescoes by timed infrared imaging analysis. *Opto-Electron. Rev.* **2015**, *23*, 102–108. [CrossRef]
17. Galli, A.; Alberghina, M.F.; Re, A.; Magrini, D.; Grifa, C.; Ponterio, R.C.; La Russa, M.F. Special Issue: Results of the II National Research project of AIAr: Archaeometric study of the frescoes by Saturnino Gatti and workshop at the church of San Panfilo in Tornimparte (AQ, Italy). *Appl. Sci.* **2023**. to be submitted.
18. Valluzzi, M.R.; Cescatti, E.; Cardani, G.; Cantini, L.; Zanzi, L.; Colla, C.; Casarin, F. Calibration of sonic pulse velocity tests for detection of variable conditions in masonry walls. *Constr. Build. Mater.* **2018**, *192*, 272–286. [CrossRef]
19. Santo, A.P.; Agostini, B.; Garzonio, C.A.; Pecchioni, E.; Salvatici, T. Decay Process of Serpentinite: The Case of the San Giovanni Baptistery (Florence, Italy) Pavement. *Appl. Sci.* **2022**, *12*, 861. [CrossRef]
20. Lanteri, L.; Calandra, S.; Briani, F.; Germinario, C.; Izzo, F.; Pagano, S.; Pelosi, C.; Santo, A.P. 3D Photogrammetric Survey, Raking Light Photography and Mapping of Degradation Phenomena of the Early Renaissance Wall Paintings by Saturnino Gatti—Case Study of the St. Panfilo Church in Tornimparte (L'Aquila, Italy). *Appl. Sci.* **2023**, *13*, 5689. [CrossRef]
21. Laureti, S.; Colantonio, C.; Burrascano, P.; Melis, M.; Calabrò, G.; Malekmohammadi, H.; Sfarra, S.; Ricci, M.; Pelosi, C. Development of integrated innovative techniques for paintings examination: The case studies of The Resurrection of Christ attributed to Andrea Mantegna and the Crucifixion of Viterbo attributed to Michelangelo's workshop. *J. Cult. Herit.* **2019**, *40*, 1–16. [CrossRef]
22. Sfarra, S.; Laureti, S.; Gargiulo, G.; Malekmohammadi, H.; Sangiovanni, M.A.; La Russa, M.; Burrascano, P.; Ricci, M. Low Thermal Conductivity Materials and Very Low Heat Power: A Demanding Challenge in the Detection of Flaws in Multi-Layer Wooden Cultural Heritage Objects Solved by Pulse-Compression Thermography Technique. *Appl. Sci.* **2020**, *10*, 4233. [CrossRef]
23. Ricci, M.; Laureti, S.; Malekmohammadi, H.; Sfarra, S.; Lanteri, L.; Colantonio, C.; Calabrò, G.; Pelosi, C. Surface and interface investigation of a 15th century wall painting using multispectral imaging and pulse-compression infrared thermography. *Coatings* **2021**, *11*, 546. [CrossRef]
24. Burgholzer, P.; Thor, M.; Gruber, J.; Mayr, G. Three-dimensional thermographic imaging using a virtual wave concept. *J. Appl. Phys.* **2017**, *121*, 105102. [CrossRef]

Disclaimer/Publisher's Note: The statements, opinions and data contained in all publications are solely those of the individual author(s) and contributor(s) and not of MDPI and/or the editor(s). MDPI and/or the editor(s) disclaim responsibility for any injury to people or property resulting from any ideas, methods, instructions or products referred to in the content.

MDPI
St. Alban-Anlage 66
4052 Basel
Switzerland
www.mdpi.com

Applied Sciences Editorial Office
E-mail: applsci@mdpi.com
www.mdpi.com/journal/applsci

Disclaimer/Publisher's Note: The statements, opinions and data contained in all publications are solely those of the individual author(s) and contributor(s) and not of MDPI and/or the editor(s). MDPI and/or the editor(s) disclaim responsibility for any injury to people or property resulting from any ideas, methods, instructions or products referred to in the content.

www.ingramcontent.com/pod-product-compliance
Lightning Source LLC
LaVergne TN
LVHW070700100526
838202LV00013B/1006